Original French Edition *Venins de serpent et envenimations* 2002
Translated from the original French by F. W. Huchzermeyer
Copyright © English Edition 2006 by IRD Editions, France

Original English Edition 2006
Printed and Published by
KRIEGER PUBLISHING COMPANY
KRIEGER DRIVE
MALABAR, FLORIDA 32950

All rights reserved. No part of this book may be reproduced in any form or by any means, electronic or mechanical, including information storate and retrieval systems without permission in writing from the publisher.

No liability is assumed with respect to the use of the information contained herein.
Printed in the United States of America.
Library of Congress Cataloging-in-Publication Data

FROM A DECLARATION OF PRINCIPLES JOINTLY ADOPTED BY A COMMITTEE OF THE AMERICAN BAR ASSOCIATION AND A COMMITTEE OF PUBLISHERS:
This publication is designed to provide accurqate and authoritative information in regard to the subject matter covered. It is sold with the understanding that the publisher is not engaged in rendering legal, accounting, or other professional service. If legal advice or other expert assistance is required, the services of a competent professional person should be sought.

Chippaux, Jean-Philippe.
 [Venins de serpent et envenimations, English]
 Snake venoms and envenomations / Jean-Philippe Chippaux ; translation by F. W. Huchzermeyer.
   p. cm.
 Includes bibliographical references and index.
 ISBN 1-57524-272-9 (hbk. : alk. paper)
  1. Venom. 2. Snakes. 3. Toxins. I. Title.

QP235.C4555   2006   2005044486
615.9'42—dc22

10 9 8 7 6 5 4 3 2

SNAKE VENOMS AND ENVENOMATIONS

Jean-Philippe Chippaux

Translation by F. W. Huchzermeyer

KRIEGER PUBLISHING COMPANY
Malabar, Florida
2006

Contents

Glossary	v
Preface	vii
Introduction	ix

PART I ZOOLOGY

1. The Zoology of Snakes	3
Palaeontology	3
Anatomy	18
The Systematics of the Snakes	24

PART II VENOMS

2. The Venom Apparatus	47
The General Evolution of the Venom Apparatus	47
The Anatomy of the Venom Apparatus	51
The Composition of the Venom	54
3. The Toxicology of the Venoms	75
Measuring Toxicity	75
Effects on Cells	78
Effects on the Nervous System	81
Effects on the Cardio-vascular System	89
Effects on Haemostasis	91
Necrotizing Activity	107
Toxicokinetics	112
The Utilization of Venoms in Research and Therapeutics	115

4. Antidotes and Immunotherapy — **125**
    Traditional Medicine — 125
    Principles and Mechanisms of Immunotherapy — 136
    Medical Treatment — 161
    Vaccination — 164

PART III ENVENOMATIONS

5. The Epidemiology of Envenomations — **169**
    The Ecology of Snakes — 169
    Human Behavior — 182
    The Epidemiology of Bites — 185

6. Clinic and Treatment of Envenomations — **211**
    Symptomatology and Clinical Pathology of Envenomations — 211
    The Treatment of Envenomations — 228
    Evaluation of the Availability of Antivenins and the Improvement of their Distribution — 239
    Prevention of Envenomations — 242

Appendix 1 Practical Measuring of the LD_{50} by the Spearman-Kärber Method — 247
Appendix 2 Symptomatic Action of Certain Plants — 249
Appendix 3 Antivenom Producers and Their Products — 257
Appendix 4 French Legislation Regarding the Keeping of Venomous Animals — 271

Bibliography — **275**

Index — **281**

Glossary

Absolute density: The real number of individuals per unit of surface or volume after an exhaustive count of all individuals
Abundance: The number of individuals belonging to one population
Action potential: Electrical wave moving along a nerve fibre and provoking a muscle contraction
Aglyphous: Maxilla with full teeth not linked to salivary or venom glands – fang-less
Alethinophidian: Group of modern shakes, ethymologically "true snakes"
Apodal: without limbs
Apparent density: Relative number of individuals per unit of surface or volume as function of a collection method or collector's activity
Autochthonous: Of local origin
Dentition: The number and distribution of teeth on the jaws; this is species specific
Diasteme: Free space on the maxilla separating two series of teeth
Duvernoy: Sero-mucous salivary gland which in the colubrids produces a toxic secretion, a precursor of the venom gland
Effector site: Precise site on a molecule where an activity is situated
Endemic: Naturally occurring in an area or region
Endothelium: Layer of cells covering the inside of blood vessels preventing leakage and setting into motion the coagulation in the case of injury
Epitope: The exact site of a molecule to which antibodies bind
Evolutionary radiation: The appearance under environmental pressure of new zoological forms from a common ancestor
Fibrillation: Repeated short muscular contraction of low amplitude
Haemostasis: Phenomenon or action aimed at stopping a hemorrhage
Incidence: Frequency of an event per period of time (usually one year); in the case of envenomations the incidence refers specifically to all bites irrespective of their severity
Jacobson's organ: Sensory organ of snakes associated with the sense of smell – vomero-nasal organ
Labial: Lip region; the labials are he scales adjacent to upper and lower lips

Lethality: The frequency of deaths caused by a given condition

Loreal: The lateral portion of the head between to eye and the nostril; the loreal shield is the shield covering this region without contact with eye or nostril

Monophyletic: Zoological groups sharing the same ancestor

Morbidity: The frequency of a disease per period of time (generally one year); in the case of envenomations the morbidity represents the symptomatic bites and thus a fraction of the incidence

Naso-vomeral organ – See Jacobson's organ

Opisthodontous: Presence of a tooth at the back of the maxilla and separated from the preceding ones by a diasteme

Opisthoglyphous: Presence of a poison fang at the back of the maxilla and separated from the preceding ones by a diasteme, back-fanged

Pleurodontous: Sentition containing different types of teeth

Population: All the individuals belonging to the same species and living in a given geographical zone

Prevalence: Number of cases diagnosed at a given moment expressing the present frequency of an event

Proterodontous: Presence of a larger tooth at the front end of the maxilla

Proteroglyphous: Tubular tooth or tooth with an open canal carried on a generally short fixed or slightly mobile maxilla, front-fanged

Purpura: Eruption of red blotches on the skin due to blood seeping out of the capillaries due to the rupture of the blood vessels or to a hemorrhage problem

Rhodopsin: Retina pigment needed for colour vision

Scolecophidian: Group of primitive still living snakes; ethymologically "worm snakes"

Solenoglyphous: Tubular fang or one with an open canal situated on a short and mobile maxilla

Symphysis: Fixed or slightly mobile articulation consisting of cartilage and elastic tissue

Territory: The space in which an animal moves and where it completes its principal activities in the rhythm of its existence (feeding, mating, egg laying or giving birth), lebensraum

Vitellogenesis: Deposition of fat reserves by the females in the ovarian follicles before ovulation

Preface

While the mythical representation of snakes in all cultures originated from an ignorance of their true nature, it was also responsible for the delay with which scientists took up the study of these very peculiar animals: Often the legends substituted observation and experimentation for their description and explication. Nonetheless, with the exception of a few countries without venomous species, poisonous snakebites are feared throughout the whole world and pose a real public health problem in the tropical regions, in particular in developing countries.

There are approximately 2700 described species of snakes of which 500 are venomous and potentially dangerous to people in varying degrees. The majority of bites result from the unexpected but not purely accidental encounter between human and snake and in most cases the latter only defends itself against what it considers to be an aggressor. Epidemiological studies aim at defining the causes and origins of these encounters. They need to be based on zoological research, which helps to identify the species involved, to study their ecology and this could lead to suggestions how to prevent these bites where possible. In addition the specificity of the venoms has been the subject of detailed biochemical and toxicological studies which have allowed us to understand the diversity of symptoms observed in cases of snakebite and to suggest specific treatments.

Every year there are 6 million snakebites in the entire world, half of them with subsequent clinical problems of varying severity. The annual number of fatal cases is estimated to be 125,000, most of them in Africa and Asia, namely in poor countries where generally treatment is inaccessible. Most snakebites happen in rural zones during agricultural activities. In contrast, in developed countries where the dangerous activities are more rare because of the mechanization of agriculture, accidental bites have appeared over the last few years which occur during recreational activities, as well as "illegitimate" bites due to handling of pet snakes, some of which are exotic and particularly dangerous. In all cases the envenomations are really difficult to take care of and to treat.

However, even beyond what one could consider a simple defense strategy, the study of the biology of snakes, of the composition and mode of action of the venoms has opened new and fascinating research perspectives regarding the molecular structure of numerous membrane receptors, the intimate mechanisms of intercellular relations or the discovery of new effective drugs against hitherto untreatable diseases.

This book offers a synthesis of recent findings and observations, allowing us to measure these riches and to understand the most significant results.

Introduction

In addition to their destructive power or the potential danger that they represent, snakes fascinate us by their appearance and by the differences which distinguish them from other vertebrates. The absence of legs, the sinuous and silent movements which seem to take place in a different dimension, the moult which is considered as a rebirth, the agile and thread-like tongue are all characteristics which man has extolled since the origin of time as evidenced by palaeolithic rock paintings.

Founding divinity of most cultures, the snake is present in all the cosmogonies, all the mythologies. It appears almost always under its two contradictory aspects, at times as virtuous power of creation and immortality, at other times monster, keeper of knowledge and bringing death, a veritable dichotomy between Good and Evil.

In the Judeo-Christian ideology the snake is the tempting demon who has brought about the perdition of man and his eternal failure, from which divine redemption alone can absolve. Without doubt the most peculiar aspect of this perception of the snake is that it is the source of Knowledge and not only a corrupter. This power, seen as primordial by people, is present in many ancient cultures. It appears in the Egyptian myth of Isis, jealous of the power of Râ, the Sun God. Isis, who embodies the feminine ideal in the Egyptian pantheon, creates a snake from clay moistened by her saliva. The snake bites the Sun God's heel and immediately the foot swells. Febrile and paralyzed by pain Râ begs for help, which the goddess gives on condition that he reveals his true identity so she can gain the knowledge and hold the same power as he. Here one sees the snake only as an instrument in the hands of the goddess and not the instigator of the plot.

In other civilizations the knowledge is transmitted spontaneously, almost plainly. The Pythians who had several famous cults in Greece, practiced divination which was intimately connected with prognosis and healing of ailments and bewitchments. Polyidos, a Cretan shepherd, learnt from a willing snake the traditional pharmacopoeia which became the corpus of the disciples of Asclepios, called Aesculapius by the Romans. The legend tells that the latter separating two snakes enlaced in copulation created the caduceus, the symbol of medicine, which goes back to the third millennium before our era. To begin with, the caduceus was the symbol of Hermes, the messenger of the Gods.

Immortality is a corollary or a consequence of therapeutic competence. In Mesopotamia, again in Crete and in Greece, but also with the Amerindian Incas, notably with Quetzalcoatl, and in Africa, it contributes to or accompanies the medical power, which the serpent exercises over the universe. In Ouidah on the coast of Dahomey, presently the Republic of Benin, the royal python is a messenger of the ancestors, the "spirits", whose recommendations and rules it transmits in the temples dedicated to the cult of the serpent. In the Hebrew world the snake is a symbol of fertility, a reputation that is found in many other societies, notably in Asia and Africa. It is difficult to explain how the serpent has acquired it, but doubtless its phallic form associated with the remarkable fecundity of certain species has played a role.

We know from the Bible that the Hebrews were decimated by snakes at Lake Suph during their exodus. Yahweh had inflicted this punishment in response to the recriminations with which they had come to him: "Fiery snakes bit the people and many died in Israel". "Fiery snakes" means viper in many passages of the Bible and the word fiery is the translation of the Hebrew "saraph" which in the middle ages became the root of "seraphin". In Greek it is translated by the word π?????? (prester), which also is the word for the species *Walterinnesia aegyptica*, an Egyptian elapid the bite of which is not only neurotoxic but also particularly painful.

The uraeus, which symbolizes the eye of Râ, becomes in Egypt the crown of Pharaos of several dynasties. The emblem representing a female Egyptian cobra (*Naja haje*) is worn as if protecting the bearer. One understands why the Egyptian cobra rather than the viper is the suicidal instrument of Cleopatra. In addition to the divine symbolism worthy of a queen of Egypt, the cobra kills rapidly without pain or physical deformation, while the viper causes a slow death with painful suffering accompanied by haemorrhages, oedemas and ugly necroses. Could the Queen of Upper and Lower Egypt in all her beauty leave to posterity on top of her failure and humiliation also the image of her disfigured body?

Introduction

In the Greco-Latin tradition the danger of serpents is clearly linked to the bite, the attributes of which are obviously distinct from those of rabies. The snake has an intrinsic deathly power, which is only occasional in the rabid carnivore and acquired by chance. This realization derives from a clinical study, which allowed Aristotle, Hippocrates and Galen to describe the different symptoms presented by bitten patients as function of different "varieties" of snake or their geographic origin. This still is limited strictly to an inventory, being descriptive rather than explicative.

In the Middle Ages, an era struggling to free itself from Antiquity, it was thought that the danger of a snake came from its fixed stare, which petrifies the victim, a veritable image of conscience.

In Africa it is often the horn-like pointed last scale of the tail that is seen as dangerous. A comparison with the sting of a scorpion, but also the behaviour of certain snakes which use their tail, although inoffensive, to lure and frighten their adversary, evidently are the origin of this widespread belief.

In Europe the tongue, a sensory organ without any toxic function or capacity to harm, disquiets and for many represents a potential danger. Here again the language analogy (snake's tongue, forked tongue) is evident, flinging out calumnies, insults, gossip and chastisements.

The discovery of the venom and its toxic properties goes back to the Renaissance. To begin with, an anatomical study allowed people to identify the poison fangs and to differentiate them from the other teeth. Then "pre-experimental" physiological studies carried out particularly by Ambroise Paré, Grévin and above all Baldo Angelo Abati in the 16th century, Francesco Redi in the 17th and Felice Fontana in the 18th were at the beginnings of the characterization of the venoms and the description of the toxic function. While Abati described the anatomy of the venom apparatus with the only error of linking the gall bladder to the poison fangs, we give credit today to Redi and Fontana for the discovery of the toxicology and the definition of its fundamental principles.

Although the complexity of the composition of the venoms was sounded by Fontana, we had to await the 19th century for the biochemical analysis and the study of the different components. From then on the study of envenomation and research of their treatment could begin. Lucien Bonaparte, the brother of Napoleon, fractionated viper venom with the help of successive precipitations with alcohol and ether. At the end of the 19th century Albert Calmette developed the serum therapy, which up to now remains the only etiological treatment of snakebites. Subsequently other researchers have separated the constituents of venoms through filtration, chromatography, electrophoresis or mass spectrometry and have described the respective properties of the fractions obtained by these methods. Very recently, molecular studies of the proteins present in the venoms, eventually modified structurally, have led to the elucidation of the most intimate cellular mechanisms, thus opening the way to entirely new domains.

Since then the work on the venoms has become oriented beyond the protection against snakebites to their utilization in biological and medical research. Heirlooms of the 20th century, but in the 21st full of promise of a fruitful future, these studies offer particularly fecund avenues to medical research and to modern therapeutics.

PART I
Zoology

CHAPTER 1

The Zoology of Snakes

Snakes, lizards, amphisbaenes and sphenodons form within the Reptilia the subclass Lepidosaurians. Snakes, lizards and amphisbaenes are characterized by a scaly skin and because of that they are placed together into a group considered to be homogenous by zoologists: the Squamata. These occur in all terrestrial surroundings and this shows good adaptation to very diverse environmental pressures.

1. Palaeontology

Snakes are the recent heirs of an ancient zoological group that for a long time dominated a remote period of prehistory. The reptiles flourished at the end of the Primary and throughout the Secondary Periods and are represented by many species that occupy all the environments. The ancestors of the Squamata are found in the lower Triassic at the beginning of the Secondary, 250 million years ago. At that time they separated from the archosaurs (dinosaurs, crocodiles and pterosaurs). The first disappeared after having widely dominated the land masses and at least partially the shallow waters. The second occupy lakes and littoral environments. The third gave rise to the birds.

1.1 The evolution of the snakes

The lizards began to diversify at the end of the Triassic (210 million years ago). During the Jurassic, in the middle of the Secondary (200-150 million years ago) the radiation – that is the diversification of forms – responded to the need to adapt to very varied nutritional regimes, notably vegetarian or carnivorous, but above all to the size of and the resistance by the prey animals for which an appropriate set of teeth and musculature of the head was required. Today the specialists agree in saying that the snakes derived from varanoid lizards, at least that the two share a common ancestor. They had thus evolved in parallel to the varanids and the heloderms, which are New World venomous lizards, to which they still are close. The primitive ancestor with whom they share the structure of the cranium and the morphology of mandible and tongue, probably was aquatic.

The particular structure of the retina, which in snakes has rods containing rhodopsine, indicates that the snakes separated from the lizards at a relatively early period (between 150 and 120 million years ago). Amongst different proposed hypotheses it is admitted today that the primitive forms were burrowing species that progressively lost their legs. The development of the retina also supports this hypothesis. In addition the primitive families of the present snakes belong to the suborder Scolecophidians, which are all burrowing. Finally there are great similarities with burrowing lizards, which also are legless and have related osteological characteristics of the skull.

The loss of legs reenforces also the need for an adaptation of the muscles and bones of the head to allow the prey to be held and to be swallowed. Consequently several bones of the head are articulated amongst each other and have become mobile to allow the mouth to be opened wide for swallowing a prey that has a larger diameter than the body of the snake. The discovery of fossils of snakes with legs has been the cause of a controversy about the evolution of snakes, as this *a priori* primitive characteristic is associated with an enlarged oral cavity in two species and a small one in the third one. In the first case it must be conceded that this primitive characteristic would have reappeared in certain species with a large oral opening or that its enlargement could have appeared independently in the two groups. The most economical hypothesis is to accept that the enlargement of the mouth preceded the loss of the legs and that all groups which had conserved external legs, have disappeared today. The existence at the same time in the middle of the Cretaceous (95 million years ago) of snakes with legs but having a small oral cavity similar to that of the Scolecophidians, which are a very primitive group, or a large mouth like the Alethinophidians, which contain the more evolved snakes, appears to confirm this hypothesis. It should also be noted that the Typhlopidae and the Boidae still have a rudimentary pelvic girdle and that some boids have horny dew-claws, which are vestiges of hind legs.

The Zoology of Snakes

The teeth followed a parallel evolution by becoming more pointed and less firmly fixed to the jaws and ending up as appendages that are implanted into bony sockets. The teeth of the Squamata serve more for holding the prey than for pulling it apart or grinding it. In fact they penetrate easily through more or less soft and thin integuments. In snakes the specialization of teeth even went further by taking the form of the venom apparatus, which will be described in another chapter.

Snakes appeared in the middle of the Cretaceous more than 100 million years ago (Fig. 1).

In that era the dinosaurs were still dominating. Next to them the snakes are much smaller animals. The land masses consisted of two continents, Laurasia which later would become North America, Europe and Asia, and Gondwana which would divide into South America, Africa and Australia (Fig. 2).

Figure 1
Stratigraphic dating of the principal snake families (after Rage, 1982)

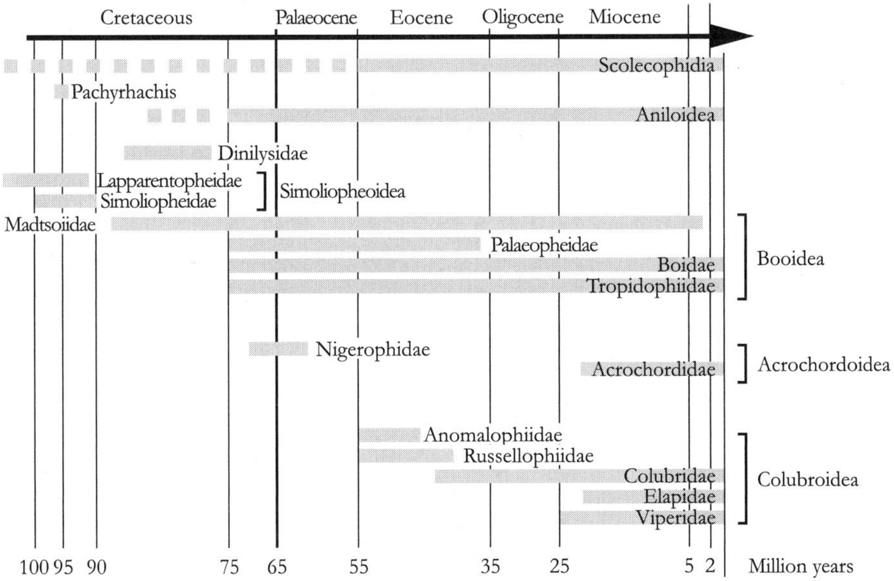

Figure 2
Locality and date of the main snake fossils

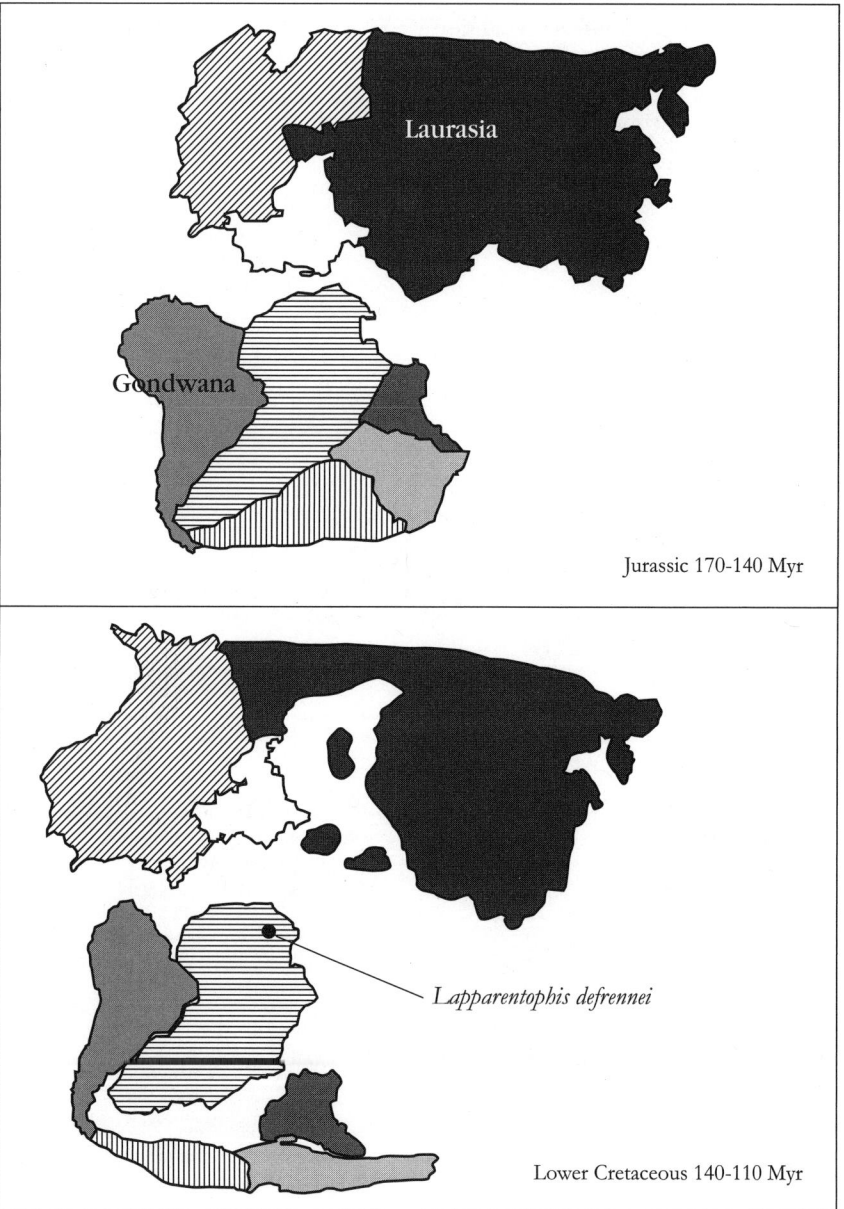

The Zoology of Snakes

(the still living genera and species are printed bold on page 25)

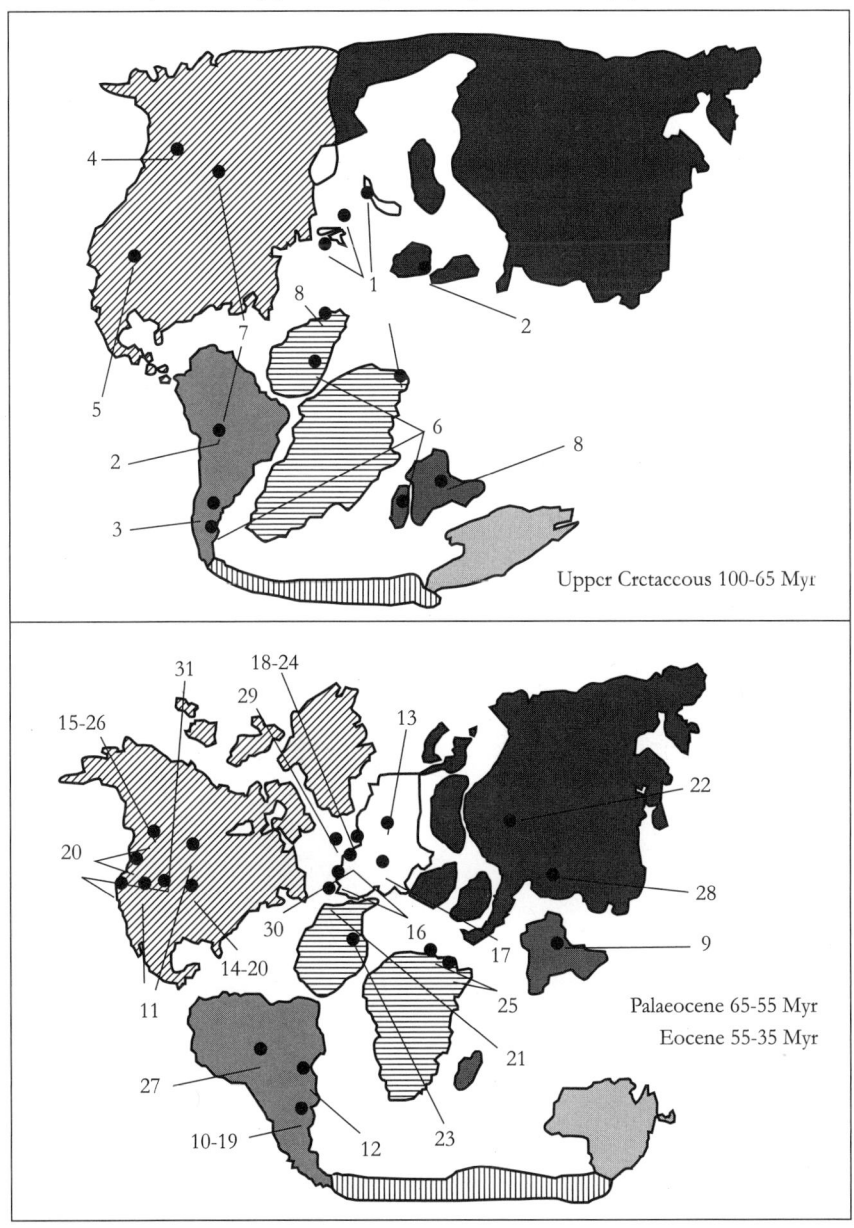

Upper Cretaceous 100-65 Myr

Palaeocene 65-55 Myr
Eocene 55-35 Myr

8 Snake Venoms and Envenomations

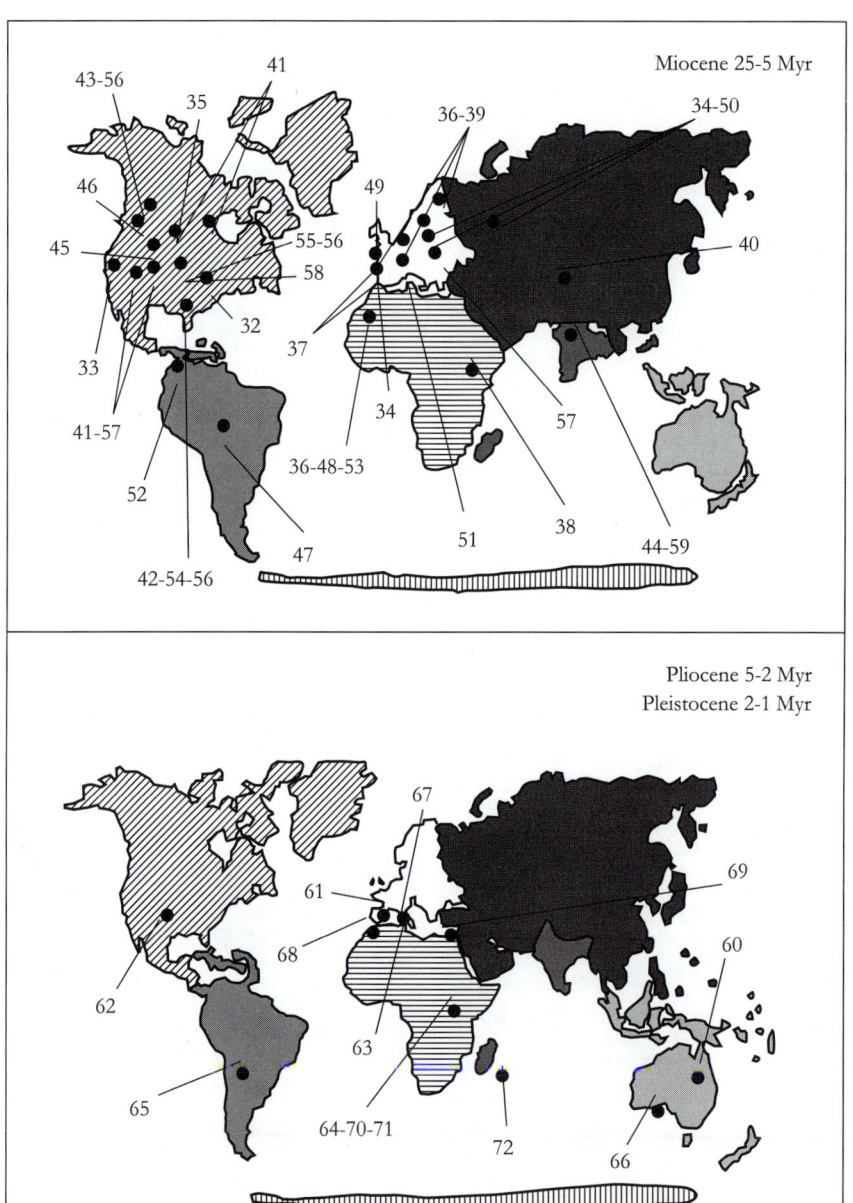

The Zoology of Snakes

Lapparentophis defrennei (Simoliopheoidea)	<100 Myr	35. *Sistrurus* indetermined (Viperidae)	22 Myr
1. *Simoliophis rochebrunei* (Simoliopheoidea)	95 Myr	36. *Vipera aspis – V. maghrebiana* (Viperidae)	22 Myr
2. *Pachyrachis problematicus* (snake with legs)	95 Myr	37. *Palaeonaja romani* (Elapidae)	20 Myr
Haasiophis terrasanctus (snake with legs)	95 Myr	38. Colubridae indetermined	20 Myr
Podophis descouensis (snake with legs)	95 Myr	39. *Natrix* sp. (Colubridae)	20 Myr
3. *Dinilysia patagonica*	80 Myr	40. *Vipera* sp. (Viperidae)	20 Myr
4. *Coniophis cosgriffi* (Aniliidae)	75 Myr	41. *Palaeoheterodon tiheni* (Colubridae)	15 Myr
5. *Coniophis precedens* ((Aniliidae)	65 Myr	*Dryinoides oxyrhachis* (Colubridae)	15 Myr
Coniophis sp. (Aniliidae)	65 Myr	42. *Sistrurus* indetermined (Viperidae)	15 Myr
6. *Madtsoia madagascariensis* (Madtsoiidae)	65 Myr	43. *Micrurus* indetermined (Elapidae)	15 Myr
Madtsoia sp. (Madtsoiidae)	65 Myr	44. *Acrochordus dehmi* (Acrochordidae)	15 Myr
7. Boidae indetermined	65 Myr	45. *Nebraskophis skinneri* (Colubridae)	12 Myr
8. Palaeopheidae indetermined	65 Myr	*Neonatrix elongata – N. magna* (Colubridae)	10 Myr
9. Boidae indetermined	65 Myr	46. *Nerodiaerythrogaster* (Colubridae)	10 Myr
10. *Coniophis* sp. (Aniliidae)	65 Myr	*Coluber constrictor* (Colubridae)	10 Myr
11. *Dunnophis* sp. (Boidae)	65 Myr	*Thamnophis* sp. (Colubridae)	10 Myr
Helegras prisciformis (Boidae)	65 Myr	47. *Eunectes* sp. (Boidae)	10 Myr
12. *Palaeophis* sp. (Palaeopheidae)	60 Myr	48. *Gongylophis* sp. (Boidae)	10 Myr
Dunnophis sp. (Boidae)	60 Myr	*Python maurus* (Boidae)	10 Myr
13. Boidae indetermined	60 Myr	Colubridae indetermined	10 Myr
14. *Palaeophis virginatus – P. littoralis* (Palaeopheidae)	60 Myr	49. *Micrurus gallicus* (Elapidae)	10 Myr
15. *Calamagras primus* (Boidae)	55 Myr	*Palaeonaja* sp. (Elapidae)	10 Myr
Dunnophis microchinus (Boidae)	55 Myr	50. *Vipera berus – V. geduyi* (Viperidae)	10 Myr
Litophis sargenti ((Boidae)	55 Myr	51. *Typhlops grivensis* (Scloecophidia)	10 Myr
16. *Russellophis* sp. (Russellophidae)	55 Myr	52. *Colombophis portai* (Aniliidae)	10 Myr
17. *Anomalophis bolcenis* (Anomalophiidae)	55 Myr	*Eunectes sirtoni* (Boidae)	10 Myr
18 Scolecophidia	55 Myr	Colubridae indetermined	10 Myr
19. *Madtsoia bai* (Madtsoiidae)	50 Myr	53. *Typhlops* sp. (Scolecophidia)	10 Myr
20. *Ogmophis voorhiensis* (Boidae)	50 Myr	54. *Micrurus* indetermined (Elapidae)	7 Myr
Boavus affinis (Boidae)	50 Myr	55. *Agkistrodon* indetermined (Viperidae)	7 Myr
Cheilophis huerfanoensis (Boidae)	50 Myr	56. *Diadophis elinorae* (Colubridae)	5 Myr
Paraepicrates brevispondylus (Boidae)	50 Myr	*Heterodonplatyrhinus* (Colubridae)	5 Myr
Tallahattaophis dunni (Boidae)	50 Myr	*Coluber s.l.* (Colubridae)	5 Myr
Dunnophis microchinus (Boidae)	50 Myr	*Lampropeltis similes – L.triangulum* (Colubridae)	5 Myr
Pterosphenus schcherti (boidae)	50 Myr	*Texasophis wilsoni* (Colubridae)	5 Myr
21. *Palaeophis* sp. (Palaeopheidae)	45 Myr	57. *Elaphe longissima* (Colubridae)	5 Myr
22. *Vialovophis zhylan* (Palaeopheidae)	45 Myr	*Natrix* sp. (Colubridae)	5 Myr
23. *Nigerophis mirus* (Nigerophidae)	45 Myr	58. *Crotalus horridus – C. viridis* (Viperidae)	5 Myr
24. *Woustersophis novus* (Nigerophidae)	45 Myr	59. *Python* sp. (Boidae)	5 Myr
25. *Gigantophis garstini* (Madtsoiidae)	40 Myr	Colubridae indetermined	5 Myr
26. *Dawsonophis wyomingensis* (Boidae)	35 Myr	*Acrochordus javanicus* (Acrochordidae)	5 Myr
Boavus idelmani (Boidae)	35 Myr	60. *Morelia* sp. (Boidae)	4 Myr
Huberophisgeorgiensis (Boidae)	35 Myr	Elapidae indetermined	4 Myr
Ogmophis compactus (Boidae)	35 Myr	61. *Palaeonaja depereti* ((Elapidae)	4 Myr
27. *Eoanilius* sp. (Aniliidae)	35 Myr	62. *Ogmophis pliocompactus* (Boidae)	3 Myr
28. *Palaeopython* sp. (Boidae)	35 Myr	63. *Eryx* sp. (Boidae)	2 Myr
Paleryx sp. (Boidae)	35 Myr	64. *Python* sp. (Boidae)	2 Myr
29. *Vectophis wardi* (Colubroidea)	35 Myr	65. Crotalinae indetermined	2 Myr
30. *Coluber carduci* (Colubridae)	32 Myr	66. *Pseudonaja* sp. (Elapidae)	2 Myr
31. *Texasophis meini* (Colubridae)	30 Myr	*Wonambi naracoortensis* (Madtsoiidae)	2 Myr
32. *Calamagras* sp. (Colubridae)	25 Myr	67. *Palaeonaja* sp. (Elapidae)	1 Myr
Pterygoboa delawarensis (Boidae)	25 Myr	68. *Naja* sp. (Elapidae)	1 Myr
Pollackophis depressusu (Viperidae)	25 Myr	69. *Palaeonaja* sp. (Elapidae)	1 Myr
33. *Proptychophis achoris* (Colubridae opithoglyphous)	25 Myr	70. *Python sebae* (Boidae)	1 Myr
Charina prebottae (Boidae)	25 Myr	71. *Bitis olduvaiensis* (Viperidae)	1 Myr
34. *Vipera sarmatica – V. ukrainica – V. aegertica* (Viperidae)	25 Myr	72. *Typhlops* sp. (Scolecophidia)	1 Myr
		Casarea sp. (Boidae)	1 Myr

The separation between the Scolecophidians and the Alethinophidians could have occurred during this era. *Lapparentophis defrennei*, the oldest fossil attributed with certainty to the order Ophidians, was found in Gondwana, more precisely in a region which is now Algeria. This species, which belongs to the super-family Simolipheoidea, which are mainly marine, is generally considered as terrestrial or by some specialists as lacustrine or living in fresh water. In the same super-family there are three related forms found on the two continents: in Laurasia on one hand several forms of snakes with legs (*Pachyrachis problematicus*, *Haasiophis terrasancticus* and *Podophis descouensis*), all found in what is now the Middle East; and on the other hand from Gondwana a *Simoliophis* sp. from the present Egypt, which was a marine snake, and *Dinilysia patagonica* from present day Argentine. This fossil shows a transition to the modern snakes or Alethinophidians. *Simoliophis rochebrunei* was found in several sites in Europe in layers dated 95 million years old. Other fossils belonging to the genus *Pachyophis* were also marine but without legs and were found in Bosnia.

The dispersion of the snakes belonging to the family of the Simoliopheidae can be explained by their aquatic way of life and the predominance of seas in that era. The other fossils of the Cretaceous are terrestrial snakes: Boidae or very closely related forms and Aniliidae and similar forms. All the fossils of terrestrial snakes from that era come from Gondwana, and this seems to indicate that they existed before the division of this continent around 120 million years ago. The expansion of the Boidae into North America could be explained by a land bridge between Gondwana and Laurasia, probably in the region of the present-day Isthmus of Panama at the end of the Cretaceous, more than 65 million years ago. This land bridge was temporary and has much enhanced the exchange of fauna between North and South America. The separation between North America and Europe was much narrower than today and the colonization of the latter by Boidae and Aniliidae could take place naturally. However, the absence of fossil Aniilidae and the rarity of fossil Boidae in the Far East, at least back to the Cretaceous, tends to show that a link between Europe and Asia was not workable and that the Ural Sea which separated the two continents until the end of the Eocene was an efficient barrier.

The Simoliopheidae disappeared during the upper Cretaceous. The Booidae persisted much longer, notably the Madtsoiidae, which became extinct about one million years ago, and the Boidae, which still are present on all continents. The Aniloidae are still in existence but limited today to Latin America and South East Asia.

The Palaeocene is a period with few reptile fossils. Particularly few Booidae, Madtsoiidae and Boidae have been found there. Numerous fossils before and after this empty period, particularly in North America and Europe, point at a reduction of the fauna during that period during which large numbers of plants and animals disappeared, including the large dinosaurs. It is supposed that the advent of a severe climate change caused a sudden temperature rise with a greenhouse effect, perhaps by darkening the atmosphere or the destruction of the ozone layer. However, most likely this extinction was selective and profited the small mammals and consequently the mammal-eating reptiles, which in turn would have benefited from it.

During the Eocene (between 55 and 35 million years ago) the northern Atlantic Ocean was formed leading to the separation of North America from Europe. The climate cooled down but remained tropical, which explains the diversification of the fauna. During the Eocene there was a considerable development of the Alethinophidians with the apparition of many species and large populations. In addition to the Boidae, the Aniloidea and the Acrochordoidea, which exist since the Cretaceous, the Colubroidea appeared including the Anomalophiidae, the Russelophiidae and above all the Colubridae. The first two families disappeared towards the end of the Eocene or during the Oligocene; the Colubridae are still present today and contain the majority of species, in number of taxa and also in population sizes. The fauna continued to be shared between North America and Europe and many genera and even species are present over this whole vast territory. The snake fauna there is regarded as Euro-American. At the end of the Eocene (35 million years ago) a glaciation took place with a severe lowering of sea levels due to the ice formation. The lowering of the sea levels allowed new land masses to appear. Europe, which until then was an archipelago, stabilized and became a true continent. The disappearance of the Ural Sea, which separated Europe from Asia, allowed the two continents to reunite. The Asian fauna colonized Europe and tended to supplant the Euro-American one, which had become very vulnerable.

The Oligocene (35 to 25 million years ago) was a very dry and cold period. The fauna again became reduced which gave another period with few fossils. Many families disappeared. The evolution of a burrowing life style because of the severe climate led to the apparition of new species. The migration of the Asian fauna towards Europe increased. The fauna became Eurasian. The Boidae still held the majority, but the Colubroidea moved in.

The Miocene began by Africa colliding with the Eurasian block. This caused a relative drying out of the Mediterranean and several intercontinental land bridges appeared at Gibraltar, in the Middle East and perhaps also Sicily. The climate changed to warmer and more humid conditions. During this period the Colubroidea exploded with the development of the Colubridae, and the Viperidae and the Elapidae began to appear. In spite of their severe reduction, the family Boidae remained, notably in the intertropical zones. This situation enhanced fauna exchanges between Africa and Eurasia, either in a south-north direction, like the proboscids, the ancestors of the elephant, which spread into Eurasia, or in the opposite direction, which very probably was the case with the snakes.

The end of the Miocene, 5 million years ago, was marked by the drying out of the Mediterranean basin during approximately one million years. This major event, the consequences of which have not yet been evaluated completely, certainly has contributed to a homogenization of fauna on both sides of the Mediterranean. The climate cooling starting with the Pliocene, completed the fauna changes by pushing back the boids, the elapids and most of the viperids as well as many colubrid species towards the tropical and equatorial regions.

It appears that in Australia the apparition of snakes, notably the elapids, occurred after the Miocene and that their introduction from Asia occurred during the Pliocene.

The present fauna therefore established itself since the Miocene, about 20 million years ago; during the Quaternary, since 1 or 2 million years ago, it has accompanied the evolution of the human species.

1.2 The origin of the present species

1.2.1 Primitive snakes or Scolecophidians

There are relatively few fossils of the very primitive group of the Scolecophidians, probably because of the fragility of their skeleton and of their small size. The oldest fossil attributed to this group comes from Belgium and goes back 55 million years. Two fossils of *Typhlops*, about 10 million years old, one coming from the south of Europe and the other one from North Africa, appear to be the oldest ones from this genus which still exists today.

1.2.2 The Aniliidae and related families

The Aniloidea are also very ancient (90 million years). They have populated South America and later North America, without doubt using a transitory land bridge between the two continents around 65 million years ago. They still are present in three genera with only non-venomous species, and live respectively in South America (*Anilius scytale*, a genus with a single species) or in Asia (*Anomochilus* and *Cylindrophis*). The family Uropeltidae consists of 7 genera that are all endemic on the Indian subcontinent. They are all burrowing species that are very close to the Aniliidae. The two families separated after the migration of the Aniloidea from Gondwana to Laurasia (65 million years ago). These two families form a transition from the Scolecophidians to the Alethinophidians, to which they belong indisputably in spite of the relative narrowness of their mouth.

1.2.3 The Boidae and related families

The Boidae are the oldest among the still existing families of the Alethinophidians. They appeared 65 million years ago within a very diversified group which itself began about a hundred years ago. The superfamily Booidea originally consisted of four different families (Madtsoiidae, Palaeopheidae, Boidae and Tropidopheidae) but now only has the two latter families; the Tropidophiidae are reduced to four single species genera, all neotropical. The Xenopeltidae also are an ancient family, which is close to the Booidea because of certain specializations.

The Booidea originated in Gondwana. From the end of the Cretaceous (about 70 million years ago) to the Oligocene (25 million years ago) they represented a particularly abundant fauna on all the land masses. The descendents are distributed between South America, Africa, Madagascar and Australia; the two latter regions have endemic forms. Profiting from a temporary Caribbean isthmus at the end of the Cretaceous, Booidea from South America entered Laurasia where they spread out. The evolution of the Booidea took place in parallel but separately on the two continents.

This group with its giant species, some possibly longer than 15 meters, disappeared from the temperate zones and established themselves in the hot regions. The last boids in Europe go back to the middle of the Miocene (15 million years ago). Today the boids have about 30 genera distributed in Latin America, Africa (including Madagascar), South East Asia and Australia. Some species are counted amongst the longest of our times, the South American boas (anaconda, *Boa constrictor*) and African pythons (*Python sebae*) and those in Asia (*Python reticulatus*). Specimens measuring up to 10 meters are still recorded. This family does not contain any venomous species.

1.2.4 The Acrochordidae

The Acrochordoidea go back to the end of the Cretaceous (50 million years ago) and are known from two fossils, one found in Niger and the other one in Belgium, both belonging to a family which no longer exists today: the Nigerophidae. They became extinct during the Paleocene (60 million years ago). The first identified arcrochordid came from the Miocene (15 million years ago) from the region of the Indo-Pakistan impact. Today the acrochordids are represented by three non-venomous species present on the South East Asian coasts and the Malayan Archipelago.

1.2.5 The Colubridae

The Colubridae appeared during the Eocene. The first formally identified colubrid fossil came from Thailand and is approximately 40 million years old. Their migration to other continents was helped by the lowering of the sea levels and the emergence of land masses at the end of the Miocene. They invaded Europe some 35 million years ago. The earliest fossils found in Europe were dated to 32 million years old. The first colubrids of North America are 30 million years old and probably arrived from Asia across the Bering Straits. Similar if not identical forms have been found in Africa in more recent strata going back 25 or 20 million years.

They would have arrived at the time when Africa joined the Eurasian block at the beginning of the Miocene. For a long time it was believed that the colonization of South America by the colubrids from North America took place in the Pliocene when the Isthmus of Panama came into being about 3 million years ago. The discovery in Columbia of a 10 million-year-old non-identified colubrid fossil suggests that their arrival could have taken place during the Miocene. It is also possible that an autochthonous group of colubrids could have appeared separately in South America. The colubrids are in fact a heterogeneous family that could contain forms of distinct origins. The exact period of the Miocene during which the colubrids colonized Australia remains unknown, but some palaeontologists think that it could have been very recently. In any case it would have been on floating rafts of vegetable matter leaving the Asian continent and carried by the currents of the sea. Whatever is the case, the colubrids never developed truly in Australia except for *Boiga irregularis* and *Boiga fusca* in New Guinea and in the north of Australia.

A 25 million-year-old fossil of an opisthoglyphous colubrid, *Proptychophis achoris*, was found in California. This would be the oldest venomous snake. Today it is accepted that the opistholyphous maxilla has appeared simultaneously in different lines of colubrids.

1.2.6 The Viperidae

The Viperidae probably originated in Asia, the only continent that is home to both the Viperinae and the Crotalinae. This family consists of venomous species only. There are no older Asian fossils than those identified in Europe, where they also appeared at the beginning of the Miocene, 23 million years ago. The first fossils of viperids in Europe belong to the genus *Vipera* that is still alive. The first crotalid fossils of the genus *Sistrurus* were discovered in North America and are 22 million years old. According to the results of biochemical and molecular analyses the American crotalids are monophyletic. They would have arrived from Asia about 25 million years ago across the Bering straits at the same time as the colubrids.

The viperids are totally absent from Australia. Some authors believe that the viperids appeared at the beginning of the Oligocene (35 million years ago or even longer). This hypothesis is based on biochemical properties of the blood and venom and also on morphological and anatomical characteristics of the different genera. The genus *Causus* is strictly African and doubtless the most primitive of the viperids. It goes back to 30 or 35 million years. The *Bitis* spp. would have arrived thereafter (30 to 25 million years ago); the two forest species (*B. gabonica* and *B. nasicornis*) would have separated much more recently, approximately 3 million years ago. The other genera appear to have established themselves about 25 to 20 million years ago. No fossils exist to corroborate this hypothesis. Curiously these genera are strictly limited to Africa, while it is generally considered that this continent was colonized at the earliest at the beginning of the Miocene (22 million years ago). In that case it would have to be admitted that the Viperidae are not monophyletic, although doubts remain.

Some of the present day species of vipers appear to have existed for 20 million years: this would be the case of the Southern European viper *Vipera aspis* while the northern species, *Vipera berus,* would be more recent (10 million years).

1.2.7 The Atractaspididae

For a long time the Atractaspididae were classified amongst the Viperidae, but they separated in fact from the other Colubroidea before the birth of the Viperidae. This family is strictly African and Near Oriental and consists of several genera of which only one has a venom apparatus of the solenoglyphous type identical to that of the Viperidae. They constitute a solid argument in favour of the concomitant apparition of the solenoglyphous maxilla in the two distinct groups.

1.2.8 The Elapidae

The Elapidae, consisting of venomous species only, would be the most recent snakes. They still are morphologically similar to the Colubridae and they would have separated from them about 20 million years ago. It is now thought that the Elapidae do not constitute a homogeneous group and that it is doubtful that they should all belong to one and the same family.

The oldest fossils of the cobras (*Palaeonaja* now extinct and *Naja* still present) go back to the beginning of the Miocene (20 million years ago) in Europe. They still live in Asia and Africa and they are the oldest known elapids. They originated in Europe, but their history on the other continents remains obscure. Probably their arrival in Africa and Asia is recent and this would explain their absence from America and Australia.

The *Micrurus* and *Micruroides* today are strictly American genera and probably are of Euro-American origin. According to the most recent works in systematics and molecular biology they would have derived from the Asian coral snakes (the genera *Calliophis* and *Maticora*), from a common Asian ancestor that would have crossed the Bering straits before its formation. The oldest fossils belonging to the genus *Micrurus* go back to the middle of the Miocene (15 million years ago) and come from North America. At the end of the Miocene they also populated Europe from where they disappeared 5 million years ago, well before the cobras that were still present in Europe during the Pleistocene (1 million years ago).

The mambas (*Dendroaspis*) are strictly African and have a maxilla that is much longer than that of the other elapids and is discretely mobile, which distinguishes them clearly from the other elapids with the exception of certain Australian elapids which have a very similar maxilla.

The origin of the Australian elapids is unknown. They also are very different from the *Micrurus* and the *Naja* morphologically as well as in the composition of their venoms. There are doubts about their Asian origins and whether they came on rafts carried by the currents like the colubrids. Genetic and immunological analyses confirm the recent apparition of the Australian elapids, which would go back to around 7 million years. The *Pseudechis* and the *Pseudonaja* would be the oldest (7 – 6 million years) ant the *Notechis* the latest to appear (3 – 2 million years). There are no fossils to confirm this hypothesis. A fossil of uncertain identification dates to 4.5 million years and another one identified as elapid was found in the south of Australia in layers going back 2.5 million years.

It is conceded that the evolutionary success of the snakes is linked to the adaptability of their mode of locomotion, which allows them to move in all the environments, and to their holding the prey with notably the apparition of the venom function. The latter will be the object of a particular chapter.

2. Anatomy

Snakes are elongate and without individualized limbs.

The skin is covered by scales. During growth the epidermis is shed in one piece in a process called moult. This occurs 10 to 15 times a year during juvenile growth and 2 to 4 times a year in adults.

The osteological, visceral and sensorial particularities are quite remarkable.

2.1 Osteology

Most of the skeleton of snakes consists of a very large number of vertebrae, 140 to 435 depending on the species, to which pairs of ribs are attached except in the tail, where these bony appendices are absent or limited to the first caudal vertebrae. The ribcage is open and without a sternum and this allows the passage of large prey with a much wider diameter than that of the bony thorax. The vertebrae are linked to each other by five articulated surfaces that allow a remarkable cohesion amongst them. In this way the spinal chord is protected while permitting the vertebral column to be flexible. Naturally this disposition is exploited during locomotion which varies very much with the nature of the ground.

Four types of movement on the ground have been described which depend on the size of the snake and on the composition of the terrain on which it takes place. The most frequent type of locomotion is creeping by lateral undulation. The undulations are wide and rapid and leave a wending and continuous spoor. The speed can reach 10 km·h^{-1}. The short and massive snakes move in a straight line much more slowly. The vertebral column continues to be straight and stretched out; successively each pair of ribs is lifted and this allows the corresponding ventral scale to hook into the substrate and to move the snake forward by a few centimeters. This movement is similar to that of millipedes. Medium size snakes move on rough ground by a telescopic or accordion-like action. The snakes holds on with the hind part of the body and lifts the head and moves it forward where it then serves as hold to pull up the rest of the body, advancing in that way. Finally on loose soil like fine sand snakes progress by lateral movement. The principle is

The Zoology of Snakes

similar to the preceding one, but the head is projected sideways and not ahead of the animal; the spoor on the ground consists of two parallel discontinuous bars. Some arboreal snakes can flatten their body and carry out a short flight between the branches.

Certain species have vestiges of a pelvic girdle: some of the bones of the pelvis and sometimes an embryo of the femur. However, there is never a sternum nor a shoulder blade, nor a trace of a forelimb.

The skull of snakes consists of a large number of small bones that are mostly articulated with one another. Without maxillar and mandibular symphyses and with the complex arrangement of the proximal articulations of these bones a snake can ingest a voluminous prey with a diameter larger than its own.

The dentition of snakes is of the "pleurodont" type (Fig. 3).

Figure 3
Dental arrangements in reptiles

Acrodontous   Pleurodontous   Thecodontous

The internal face of the jaw is oblique forming a bony bulge to which the teeth are attached. They are replaced continuously in successive waves affecting alternatingly one of two teeth. The replacement teeth grow in the gingival mucosa and several generations of teeth of increasing size can be seen simultaneously. The poison fangs are of the same type of dentition and consequently a fang that is pulled out either accidentally or intentionally, is replaced almost immediately and several replacement fangs are ready to take the place of the preceding one when necessary.

The teeth are thin, pointed and curved inwards. In the boids the teeth decrease in size from front to back. In the other snakes with the exception of a few rare fangless colubrids, the teeth increase in size from front to back. Their role is essentially to hold the prey and to pierce the integument.

The anatomy of the venom apparatus is described further below.

2.2 The internal organs

The general layout of the internal organs in snakes is very particular. It is adapted to the elongated shape of the animal. The viscera are elongated and arranged one behind the other; some have atrophied.

The heart has two auricles, which are asymmetrical, but only one ventricle like in most reptiles. An incomplete membrane divides the ventricle and prevents the mixing of oxygenated blood coming from the lungs with the blood returning from the body. The position of this organ depends on the way of life of the snake and allows the heart to compensate for the effects of weight. Generally the heart of snakes lies at the anterior third for an equilibrated distribution of the blood in any of the positions adopted by the animal. In the tree-living snakes the heart lies very close to the head to facilitate the supply to the brain during climbing. In contrast, in aquatic snakes the heart is in the center of the body to compensate for the hydrostatic pressure.

The left lung is missing or atrophied. The right lung takes about one quarter of the length of the snake and has few alveolae. The respiratory exchanges are relatively weak which is partially explained by the low metabolic rate of snakes. The modest energy requirements and the oxygen reserves are the reason for the prolonged resistance to anoxia by snakes: elapids can suspend their respiration for 30 minutes and viperids for 95 minutes.

Stools and urine are excreted through one opening: the cloaca, which results from an anastomosis of the ureters with the intestine. The kidneys are elongate and of the metanephros type, the filtration of the urine occuring at the end of the nephron. This is a precursor system to the glomerular filtration of the mammals. There is no bladder.

2.3 The reproductive organs

The testicles and ovaries are elongated, paired and generally situated one behind the other. The males have paired and symmetrical copulatory organs, the hemipenises, which in their resting position are within the base of the tail. These hemipenises have a cavernous body which allows their erection and on the surface a complex ornamentation consisting of spines and protuberances which help to hook the organ during copulation.

The female has hemiclitorises that correspond to the hemipenises. The ovaries are strung out in the abdominal cavity.

Ovulation is generally seasonal. It is preceded by a long period of preparation and the deposition of energy and nutritional reserves which are necessary for the development of the embryo: vitellogenesis. During this period relatively important anatomic modifications take place. Most of the species are oviparous but vipers are ovoviviparous and give birth to little ones that have undergone their development in the eggs entirely within the oviducts of the female.

2.4 The sensory organs

Snakes have no external ears. The middle ear is rudimentary and only represented by the columella that replaces the three ossicles which transmit the vibrations of the tympanum to the internal ear in mammals (Fig. 4).

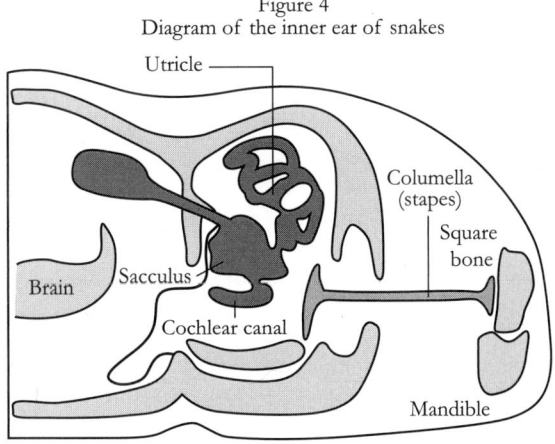

Figure 4
Diagram of the inner ear of snakes

2.4.1 Hearing

The columella is the equivalent of the anvil. There is no Eustachian tube nor a tympanum. The inner ear is organized in a particular way. The semi-circular canals, utriculus and sacculus are developed normally and serve the functions of equilibrium. The cochlear canal with the basilar membrane paradoxically is well developed in spite of the absence of outer and inner ears and forms the only true hearing organ. The basilar membrane senses the vibrations that are transmitted not by the organs of the middle ear, but by the quadrate bone which transmits the vibrations from the head to the internal ear via the columella. Consequently the snake does not hear in the classical sense but analyzes the vibrations received by the whole body and relayed to the internal ear.

2.4.2 Vision

The eyes of the Scolecophidians have atrophied. All other snakes have a good sense of vision. They have no eyelids and the eye is protected by a fixed transparent scale.

The field of vision is relatively large, in the order of 120 to 140 degrees and most of the species have a binocular vision that is necessary for judging distances and evaluating uneven terrain. The retina of the Scolecophidians only contains rods for black and white vision in semidarkness. The retina of the boids has cones for daytime colour vision in addition to the rods. The more developed species, colubrids, cobras and vipers, have a relatively complex modification of retina types with rods and cones of different shapes which probably enhance the visual acuity in adaptation to their behaviour.

For visual accommodation and the projection of a sharp image onto the retina the lens can be brought forward or retracted by the contraction of the ciliary muscles. The pupil is round or elliptical and can be dilated or closed depending on light intensity.

2.4.3 Olfaction

Not much is known about the olfactory capabilities of snakes. In any case, the size of the olfactory bulbs and peduncles of the telencephalon of snakes indicates that this sensorial function is particularly well developed. Associated is a sense organ

The Zoology of Snakes

specific to snakes, the vomeronasal or Jacobson's organ, which is situated below the nasal fossa with an opening in the vault of the palate (Fig. 5).

The ends of the forked tongue of the snake fit into it and bring adhering substances into contact with the sensory cells lining the organ. The movements of the tongue are very rapid and this allows volatile substances present in the environment to be analyzed immediately. This demonstrates the importance of this sensorial function in the capture of prey, in the preparations for mating and in the mechanisms of defense.

2.4.4 Thermosensitive receptors

The crotalids and some species of boids have heat sensitive pits. In crotalids a pair of deep pits in the loreum allows the detection of the radiation of a living being, prey or predator, and to follow its displacements. In the boids several labial pits fulfill this function.

The loreal pit of the crotalids consists of a cavity that is closed in its middle part by a membrane which is heat sensitive and linked to the trigeminus nerve, the facial sensory nerve of most vertebrates (Fig. 6).

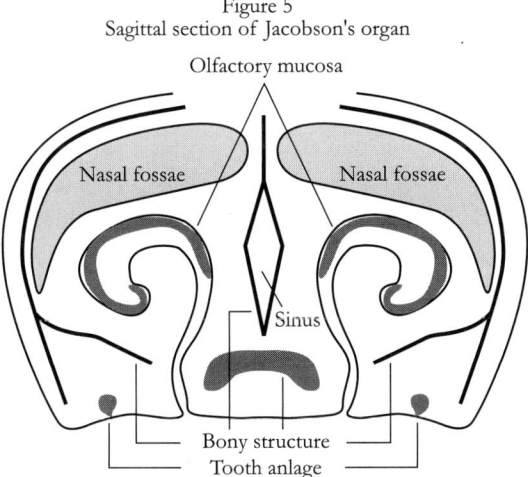

Figure 5
Sagittal section of Jacobson's organ

Figure 6
Diagram of the thermo-sensitive pit of a crotalid
(after Newman and Hartline, 1982)

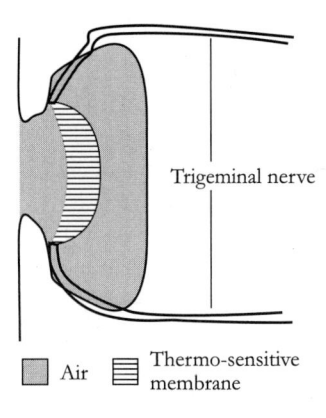

The membrane absorbs infrared radiation and transmits the excitation to the nerve. The sensitivity is in the order of 0.003° C. In boids the epithelial lining of the pit consists of very superficial sensory receptors and this increases their sensitivity. The number of receptors determines the precision of localizing the source of heat. The angle of perception is in the order of 5°.

Thermal detection optimizes vision, notably during prey capture. Thus a moving prey animal emitting heat is perceived both by the eye and the thermosensitive pit and this allows it to be distinguished from a fixed object, even a heat source, or from a moving but cold one.

3. The Systematics of the Snakes

The reptiles to which the snakes belong, are ectotherm vertebrates whose temperature depends on that of the environment. They are amniotes whose egg has a vitelline membrane that encloses the embryo and three extra-embryonal membranes: the amnion, the chorion and the allantois. They have a direct development without a larval stage.

The snakes or ophidians together with the lizards and the amphisbaenes belong to the order Squamata. They are now classified in two suborders (Table I).

3.1 Scolecophidians

They probably are the most primitive of the snakes and are of small size. They are called "minute" snakes from the Latin "*minutus*", which refers to their size and not

Table I
The classification of snakes (after McDiarmid, *in* Seigel *et al.*, 1987)

Family	Genera	Species	Distribution
Scolecophidians			
Leptotyphlopidae	2	87	Cosmopolitan
Anomalepididae	4	15	Latin America
Typhlopidae	6	203	Cosmopolitan
Alethinophidians			
Anomochilidae	1	2	Sumatra
Uropeltidae	8	47	South of India and Sri Lanka
Cylindrophidae	1	8	South East Asia
Aniliidae	1	1	Latin America
Xenopeltidae	1	2	South East Asia
Loxocemidae	1	1	Central America
Boidae	16	67	Cosmopolitan
Bolyeriidae	2	2	Southern Indian Ocean
Tropidophiidae	4	21	Central America and Caribbean
Acrochordidae	1	3	South East Asia, Australasia
Colubridae	255	1 405	Cosmopolitan
Atractaspididae	6	55	Africa, Middle East
Elapidae	60	298	Cosmopolitan (tropical)
Viperidae	33	235	Cosmopolitan

to the danger of their bite. In fact, they have no poison apparatus at all. These snakes are blind and of burrowing habits. Most often the eye is reduced to a pigmented spot under one or several scales of the head. The body is covered in small shiny scales, identical on back and belly and also on body and tail. Most of the species have a vestigial pelvis but no trace of lower limbs. This suborder consists of three families: the Leptotyphlopidae, the Anomalepididae and the Typhlopidae.

3.2 Alethinophidians

This group contains the modern snakes whose classical morphology is well known. The eyes are complete with retinas consisting of rods and cones.

The ventral surface is covered with large scutes that differ clearly from the small triangular or lozenged scales covering the back.

This suborder consists of fourteen families and the main ones of these are: the Acrochordidae, the Aniliidae, the Uropeltidae, the Xenopeltidae, the Boidae, the Colubridae, the Atractaspididae, the Elapidae and the Viperidae. Only the last four families, which probably are the most recent ones, have venomous species. It is mainly those, of course, with which we are dealing here.

3.2.1 Colubridae

The general aspect of the colubrids is the classical one of snakes.

The body is long and thin. The size of the tail varies but it is generally filiform. The head is rounded and not much set off against the body except for a few species with a well-defined neck.

The pelvic girdle is totally absent.

The maxilla, the palate, the pterygoid and the dentary carry all the teeth. In contrast, the premaxilla never has any. The maxilla is long and can have one or several diastemata. If there are poison fangs they always are in posterior position (opisthoglyphous).

The head is covered by head shields (Fig. 7).

The colubrids have 250 genera and 1400 species, most of them are not dangerous to man. This family is far from homogeneous and all attempts at classification have been fruitless.

The Opithoglypha have back fangs that cannot inoculate large volumes of venom, and only represent a relative danger. Some species have a particularly toxic venom and fangs positioned sufficiently forward and long to be able to inflict severe and even deadly bites, when the animal is handled intentionally.

Figure 7
Scales on the head of a colubrid

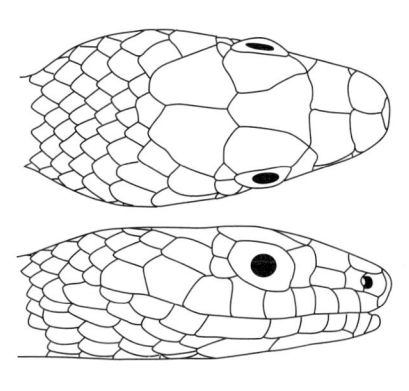

Colubrid bites occur often: the abundance of species belonging to this family, their wide distribution and for several decades the fascination of naturalists for these snakes explain this phenomenon. The literature mentions a number of accidents, some of them fatal, caused by species belonging to the genera *Dispholidus, Thelotornis, Boiga*, to which could be added *Toxicodryas* which recently was separated from them, *Malpolon, Rhabdophis, Tachymenis, Philodryas, Oxybelis* and *Clelia*. Some aglyphous colubrids (but generally opithodontous ones like *Hydrodynates gigas* or *Waglerophis merremii*) have also been accused of having caused serious envenomations. The toxic saliva produced by Duvernoy's gland can be introduced into the body through the bite wound. However, in spite of frequent bites, no opisthoglyphous species of the genera *Psammophis* and *Crotaphopeltis* appear to have caused any serious envenomations.

Dispholidus. This African genus contains a single species. *Dispholidus typus* is an arboreal species of the savannah or the open forest. Several fatal bites have been described, always during intentional handling.

The venom contains a glycoprotein belonging to the serine-proteases, venobin, which activates prothrombin. The symptoms are essentially haematological. The inflammatory syndrome is discrete and there is no characteristic neurological sign. A specific antivenin is produced in South Africa.

Philodryas. This South American genus consists of 18 terrestrial and ubiquitous species of which the most important ones are: *Philodryas olfersii, P. aestiva, P. nattereri, P. patagoniensis, P. psammophidea* and *P. viridissima*.

The venom of *P. olfersii* contains a myotoxin and a fibrinolytic enzyme that catalyzes fibrinogen and fibrin. The envenomations by *Philodryas* cause a haemorrhagic syndome that is very similar to the one provoked by the *Bothrops* spp., which have a similar geographic distribution and this can cause a problem with the differential diagnosis. In any case, the antivenin prepared against the venom of *Bothrops* has no effect on an envenomation by *Philodryas*. While occurring frequently the envenomations are generally not severe, although *P. olfersii* and *P. patagoniensis* have been made responsible for several deaths following poorly documented envenomations.

Rhabdophis. Nineteen species belong to this Asian genus and are distributed from Myanmar to the Philippines. The principal species are: *Rhabdophis subminatus, R. chrysargos, R. himalayanus, R. nigrocinctus, R. nuchalis* and *R. tigrinus*.

The venom of *R. subminatus* has a high toxicity (50% lethal dose of 25 µg per mouse of 20 g). It has an intense proteolytic activity and an activator of blood coagulation factor X has been described. Platelet aggregation is strong, but there is no fibrinolytic or fibrinogenolytic action. This species causes serious and sometimes fatal envenomations, although the capacity of the venom glands is relatively low. The poison fangs are particularly long. The envenomation is characterized by a haemorrhagic syndrome, at least locally, with permanent bleeding from the bite wound.

Tachymenis. This is a South American genus with five species spread over Argentina, Chile, Peru and Bolivia. One of these, *Tachymenis peruviana*, has inflicted severe envenomations. Its venom has not been studied in detail.

Thelotornis. This is an arboreal African genus with two species: one eastern and southern, *Thelotornis capensis*, and the second one central and western, *T. kirtlandii*. These are very thin snakes, called liana snakes, perfectly mimicking their environment.

The venom contains a prothrombin and a factor X activator. It also has a fibrinolytic action to which is added, according to some authors, a thrombin-like action. The symptomatology is marked by a severe and sometimes fatal haemorrhagic syndrome. There is no antivenin against this species. In view of the absence of a cross-neutralization with the venom of *Dispholidus typus* and even more so with those of the viperids, there is no effective serum therapy.

Boiga. This is an Asian and Australian genus with the principal species: *Boiga dendrophila, B. cynodon* and *B. irregularis*. The venom has a phospholipase action and several other enzymatic actions of lesser toxicity (phosphodiesterase, amino-acid-oxidase) or moderately toxic and therefore of little danger (cholinesterase-like, proteolytic). The secretions of *B. dendrophila* cause haemorrhage.

Toxicodryas. This is a arboreal African genus with two species of which the main one is *Toxicodryas blandingii*.

The venom contains a neurotoxin of light molecular weight (7 to 9 kDa) with an important synaptic tropism that in its effects are very similar to that of the elapids. There is no biochemical or immunological relationship between the neurotoxin of *Toxicodryas* and those of the elapids. A cardio-vascular collapse is often seen. However, there does not appear to be any haemorrhagic problem in spite of an intense proteolytic action of the venom.

Malpolon. This is a genus of the Mediterranean region, which is represented in the south of Europe by *Malpolon monspessulanus*, and in Saharan Africa by *M. moilensis*.

The envonomation is characterized by a local to regional inflammatory syndrome (pain, oedema, lymphangitis) and in some cases by neurological symptoms (paraesthesia, difficult swallowing, ptosis and dyspnoea). A paralysis of the cobra type is possible but seems to be exceptional. A post-synaptic neurotoxin without structural relationship to the toxins of the elapids is said to be the cause. In the mouse a particular toxin provokes a fatal pulmonary haemorrhage.

Oxybelis. This arboreal South American genus has four species of which the most widely occurring ones are: *Oxybelis fulgidus* and *O. aeneus*.

The composition of the venom is unknown, but it appears that the few reported human envenomations were benign.

Clelia. This genus is present in all of Latin America and its only species is forest-dwelling, *Clelia clelia*.

Some envenomations by *Clelia* have been described. They were always limited to local symptoms without any systemic spread: oedema, ecchymoses, discrete focal haemorrhages and sometimes a limited necrosis.

3.2.2 Atractaspididae

This family is a particular case amongst the snakes. It contains several genera of which only one has a solenoglyphous venom apparatus. The other ones are either aglyphous or opithoglyphous, which perfectly illustrates the adaptive opisthodontous route. Successively classified with the Viperidae, then the Colubridae, the genus *Atractaspis* together with some African genera of minor importance, has recently been established as belonging to a distinct family. It is endemic in Africa and the Middle East and mainly contains primitive burrowing species.

Atractaspis (burrowing adder). This genus has 17 species of which the most important ones are: *Atractaspis bibronii, A. aterima, A. dahomeyensis, A. engaddensis, A. irregularis, A. microlepidota*.

The venom of the *Atractapis* is composed of enzymes and toxins that are strongly cardiotropic, the sarafotoxins. The characteristic symptoms are an inflammation and often a localized necrosis. Severe and sometimes fatal problems of the heart rhythm often complicate the envenomation. There is no antivenin against the venom of these species.

3.2.3 Elapidae

This is a heterogenous family and probably derives from several distinct ancestors, which in parallel but separately acquired a proteroglyphous dentition. It contains 60 genera and about 300 species including the marine snakes. They look just like the Colubridae. The head scales consist of large shields (cf. Fig. 7).

In Africa several genera share the different ecological niches. Some of them are endemic, like the arboreal *Dendroaspis* and *Pseudohaje*, *Boulengerina*, which is aquatic, *Elapsoidea, Aspidelaps, Paranaja* and *Walterinnesia*, which are burrowing. In contrast, the genus *Naja* that has many African species occurs in Africa and Asia, as well as in Europe during the Miocene.

In America the coral snakes (*Micrurus* and *Micruroides*) are burrowing and snake eaters.

In Asia, in addition to *Naja* and *Bungarus*, there are several genera related to the coral snakes, *Maticora* and *Calliophis*, as well as the king cobra *Ophiophagus hannah*.

In Australia there are many and particularly diversified elapids. As in Africa there are terrestrial, burrowing, aquatic and arboreal species. The proteroglyphous sea snakes form a particular group, if only because of the biotope in which they live.

The symptoms presented in the case of an envenomation by elapids are typically those of the cobra syndrome that manifests itself by a locomotor paralysis equally affecting the respiratory muscles and leading towards death by asphyxia. In some cases, specified below, a particular symptomatology is observed, either particular neurological signs but distinct from the classical cobra syndrome, or symptoms of inflammation, necrosis and haemorrhage.

Acanthophis (death adder). This Australian genus also occurs in Papua New Guinea and in Indonesia and contains three species, the best known of which is *Acanthophis antarcticus*.

The venom is strongly neurotoxic because of the presence of an α-neurotoxin. A specific antivenin is produced in Australia.

Boulengerina (water cobra). This African genus is strictly aquatic and piscivorous and has three species of which the main one is *Boulengerina annulata*. Bites are rare and usually occur during fishing activities, like cleaning the nets. The venom contains an α-neurotoxin. There is no specific antivenin, but the polyvalent antivenins prepared for Central Africa appear to be effective.

Bungarus (krait). This Asian genus consists of 12 species of which the principal ones are: *Bungarus candidus, B. fasciatus, B. caeruleus, B. multicinctus*. They are nocturnal terrestrial snakes, living close to human communities. *B. candidus* is essentially responsible for bites in the house inflicted during the victim's sleep, probably as response to a defensive movement. *B. fasciatus* frequents rice paddies where at night it causes quite a large number of bites.

The venom is strongly neurotoxic because of a pre-synaptic β-neurotoxin in addition to which certain species also have an α-neurotoxin. The locomotor paralysis can develop into a fatal asphyxia. Polyvalent or specific antivenins are produced in India, China and Thailand.

Dendroaspis (mamba). This strictly African arboreal genus consists of four species: the green mambas *Dendroaspis angusticeps* in East and Southern Africa, *D. jamesoni* in Central Africa, *D. viridis* in West Africa, all in the forests, and the black mamba *D. polylepis* of the savannahs in intertropical Africa.

The venom contains phospholipases that in principle have no pre-synaptic toxicity, and several types of blocking or facilitating neurotoxins which translate into a complex and paradoxical clinical picture. The dendrotoxins and the fasciculins enhance the nervous transmissions, while the α-neurotoxins block this transmission. The delay and the duration of the action of these toxins differ and this leads to symptoms with a capricious development. The bite of a mamba can be inflamed and painful. It is above all marked by neurological problems of the muscarin-like type (sweating abundantly, lacrimation, disturbed vision, abdominal pain, diarrhea and vomiting), which precede the cobra syndrome. The polyvalent African antivenin produced in France is very effective.

Hemachatus (ringhals, spitting cobra). This South African genus is monospecific, *Hemachatus haemachatus*.

The venom can be sprayed from a distance on the aggressor or the prey. When it is close it aims at the eyes of the victim like the spitting cobras, causing a very painful and blinding inflammation of the conjunctivae if not treated appropriately. However, the bite with the inoculation of venom remains the principal means of defense or of prey capture.

The venom contains like that of the *Naja* an α-neurotoxin, which causes a locomotor paralysis which can develop towards death by asphyxia. An effective antivenin is produced in South Africa; the one produced in Europe appears to have a certain degree of cross reactivity.

The Zoology of Snakes

Micrurus (coral snake). This American genus contains 64 species of which the principal ones are: *Micrurus corallinus, M. fulvius, M. hemprichii, M. lemniscatus, M. narduccii, M. nigrocinctus* and *M. surinamensis*.

The venom contains a post-synaptic α-neurotoxin that makes it strongly neurotoxic. Monovalent and polyvalent antivenins are produced in the United States, Costa Rica, Mexico and Brazil.

Naja (cobra). This genus contains 18 species spread over Africa and Asia. The principal species are in Africa *Naja haje, N. nigricollis, N. mossambica, N. melanoleuca* and in Asia *Naja naja, N. kaouthia, N. oxiana, N. sputatrix*.

The venom of all these species consists of phospholipases generally without presynaptic toxicicity, of cardiotoxins and of α-neurotoxins, which are short or long depending on the species. Most have a high neurotoxicity. Some (*N. nigricollis, N. mossambica* which can spit) cause inflammation at the site of the bite, which can develop into a restricted necrosis. A polyvalent antivenin for Africa is produced in France; monovalent or polyvalent antivenins for Asia are produced in India, Thailand, China, Vietnam and the Philippines. The cross reactivity amongst the species is mediocre, particularly between the *Naja* spp. from Asia and those from Africa.

Notechis (tiger snake). This Australian genus contains two species of which the most important one is *Notechis scutatus*.

The envenomation is characterized by the combination of a locomotor paralysis, muscle spasms, a haemorrhagic syndrome and sometimes renal insufficiency. A specific antivenin is produced in Australia.

Ophiophagus (king cobra). This is an Asian monospecific genus, *Ophiophagus hannah*, which is morphologically very close to the *Naja* spp.

The venom is strongly neurotoxic. A monovalent antivenin is produced in Thailand and a polyvalent one in China.

Oxyuranus (taipan). This Australasian genus has two species of which *Oxyuranus scutellatus* has the most wide-spread distribution in Australia, Papua New Guinea and Indonesia.

The taipans are amongst the most dangerous snakes of the world. Before the Australian specific antivenin became available, more than 80% of the bites were lethal. The envenomation combines most of the clinical signs found in snake envenomations: paralysis, muscle necrosis, a haemorrhagic syndrome and acute renal insufficiency. The venom contains the β-neurotoxin with presynaptic toxicicty taipoxin, a cardiotropic toxin taicatoxin and a prothrombin activating enzyme. The symptomatology is complex combining neurological problems (the classical cobra syndrome), disturbance of the heart rhythm and a haemorrhagic syndrome. The bite is painless and the local inflammation is negligible. Bleeding occurs in a third of the envenomations but responds well and quickly to the antivenin treatment. In contrast, the cobra syndrome responds poorly to the antivenin, once it is established and artificial respiration is often necessary. A specific antivenom is produced in Australia, which has to be given early (one to two hours) for optimal efficacy.

Pseudechis (mulga snake). This is an Australian genus, but one of the six known species, *P. papuanus*, also occurs in Papua New Guinea and Indonesia; a second one, *P. australis*, also occurs in Indonesia.

The venom is less toxic than that of the other Australian elapids. However, because of the exceptional capacity of the gland, particularly in *P. australis*, the envenomation can be severe and even fatal. The symptoms are relatively indistinct: muscle necrosis and renal insufficiency are the main signs that are seen. A specific antivenin against *P. australis* and *P. butleri* is produced in Australia. The antivenin produced against the venoms of *Notechis* has a cross reactivity against the venoms of the genus *Pseudechis*.

Pseudonaja (pseudonaja). Originally this is a strictly Australian genus consisting of seven species of which one, *Pseudonaja textilis*, has been introduced by man to Papua New Guinea.

The envenomation by the *Pseudonaja* spp. is dominated by haemorrhagic problems and renal insufficiency. Locomotor paralysis and heart problems with cardiovascular collapse are sometimes seen. A specific antivenin is produced in Australia.

Sea snakes. Five genera of sea snakes have been described and some of these have many species (33 species alone for the genus *Hydrophis*). These snakes are strictly aquatic, even if they happen to leave the water to come onto the beaches of the Pacific, and cause very few bites. Accidents mainly occur when the snake is handled and bites in defense. This happens when traditional fishermen empty out their nets and when divers try to catch the snake. Contrary to a persistent reputation, the sea snakes can bite and inoculate their venom without difficulty. The latter contains neurotoxins that lead to locomotor paralysis and asphyxia which can be fatal. There is a great risk of drowning when the bite is received in the water.

The main species are in order of their frequency or distribution:

- *Aipysurus eydouxii, A. laevis;*
- *Enhydrina schistosa;*
- *Hydrophis ornatus, H. coggeri, H. cyanocinctus, H. gracilis;*
- *Lapemis curtus;*
- *Laticauda colubrina, L. laticauda, L. semifasciata;*
- *Pelamis platurus.*

3.2.4 Viperidae

Although they have a loreal sensory pit, the Crotales have now been placed into the family Viperidae. Except for the genera *Causus* and *Azemiops*, which consist each of a single-genus subfamily, and even single species in the case of *Azemiops*, the whole family appears relatively homogeneous. In a general way the Viperidae seem to show a particularly well-developed adaptability. They have colonized many environments, some of them as hostile as deserts or boreal regions. Generally they are terrestrial, but there are some arboreal species on all continents. The Viperinae are found only in the Old World and the Crotalinae are found in Asia and in America.

This family has 33 genera and 235 species.

3.2.4.1 Causinae

This single-genus subfamily is relatively primitive and strictly African. The head is covered with head shields that are identical with those of the Colubridae (cf. Fig. 7). In most of the species the venom gland is elongated. It occupies the first quarter or fifth of the body under the skin. The gland is not surrounded by skeletal muscle as in the other viperids.

Causus (night adder). This genus has six species of which the most widely spread ones are *Causus rhombeatus* and *C. maculatus*. These species are ubiquitous and commensal. They are found in gardens and plantations where they eat toads.

The venom is not very toxic for mammals and the envenomations, though painful, are generally not very serious except for an oedema sometimes accompanied by a temporary fever. There is no antivenin.

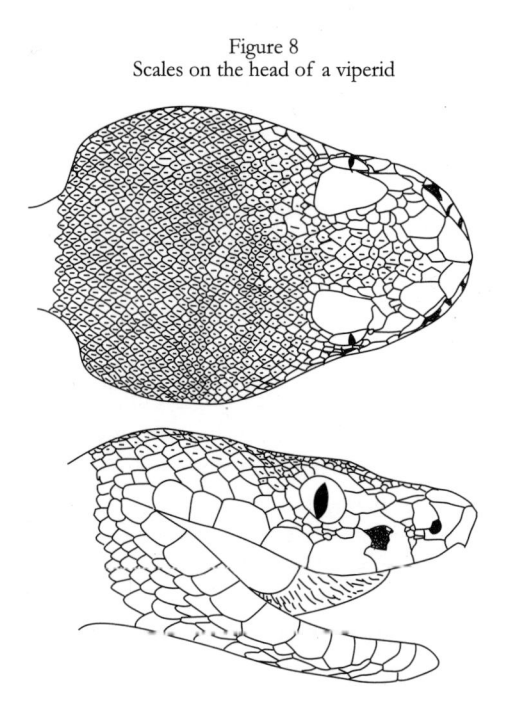

Figure 8
Scales on the head of a viperid

3.2.4.2 Viperinae

Their body is robust, the tail short and the head triangular with a distinct neck. The head is covered in small scales that are identical to those which cover the back (Fig. 8).

In Europe there is only one genus, *Vipera*, and it is widely distributed. It is part of a Eurasian group which stretches to South East Asia and the Arabian Peninsula. There one finds in addition to *Vipera*, the genera *Daboia*, *Macrovipera*, *Montivipera*, *Pseudocerastes* and *Eristocophis*.

The genera *Bitis*, *Atheris* and *Adenorhinos* are African.

The genus *Cerastes* is endemic in North Africa and the Middle East and is of uncertain origin.

The monophyletic genus *Echis* extends from West Africa to India and Sri Lanka.

Atheris (bush viper). This arboreal African genus contains 10 species of which the most widely distributed ones are: *Atheris sqamigera* and *A. chlorechis*.

The venom causes inflammation and haemorrhage. Several deaths have been reported after an envenomation by *Atheris squamigera*. There is no specific antivenin.

Bitis. This African genus has 16 terrestrial species, amongst them *Bitis arietans* (puff adder), *B. gabonica* (Gaboon viper) and *B. nasicornis* (rhinoceros viper). These thick vipers cause 5% of the bites in the savannah and 10% in the forest.

The venom causes strong inflammation, necrosis and haemorrhage, A thrombin-like enzyme has been isolated from the venom of *B. arietans* and from that of *B. gabonica*. However, the venom of *B.arietans* causes more necrosis and that of *B. gabonica* more haemorrhage. In addition it contains an inhibitor of the angiotensin conversion enzyme, which causes a cardio-vascular shock. The African polyvalent antivenin produced in France is very effective.

Cerastes (horned viper). This genus lives in the deserts of North Africa and the Near East and has three species, the two principal ones of which are *Cerastes cerastes* and *Cerastes vipera*. The venom causes strong inflammation and necrosis. A haemorrhagic syndrome is often observed. There are several polyvalent antivenins produced in France, in Algeria and in Saudi Arabia.

Daboia (Russel's viper). This Asian genus was monospecific (*Daboia russelii*) but three other species have been added: *D. palaestinae* (Palestine viper) in the Near East, *D. deserti* (Desert viper) in Tunisia and North-West Lybia and *D. mauritanica* in North Morocco and North Algeria. They used to be in the genus *Vipera* from which they have been separated recently.

D. russelii, the most important species of this genus, has a discontinuous distribution from India to Taiwan and Indonesia. This species is diurnal and causes a large number of bites amongst agricultural workers, particularly in the rice paddies where it abounds. In China less than 5% of the bites are attributed to *D. russelii siamensis*, a subspecies that also occurs in Thailand, where the incidence is around 15% and in Myanmar, where it is responsible for 70% of the bites. A similar frequency of bites is observed in certain parts of India with *D. russelii russelii*, which perhaps is explained by the behavior of the human population which is quite similar in those two countries. In Sri Lanka the frequency of bites by *D. russelii pulchella* does not exceed 5%.

Five subspecies of *D. russelii* have been described with a corresponding geographical distribution. The composition of the venom and with it the symptomatology of the envenomations varies with the geographical regions. In all the countries the venom provokes inflammatory and haemorrhagic syndromes, which frequently lead to renal insufficiency. Thrombo-embolic accidents, particularly involving the brain or the hypophysis, are particular for *D. russelii siamensis* living in the south of India and in Myanmar. Neurotoxic and muscular problems, probably due to the presence of phospholipase A_2 in the venom, are encountered in Sri Lanka and in the south of India with *D. russelii pulchella*. The causes of the renal insufficiency observed in Myanmar appear to be multifactorial. The venom of this subspecies (*D. russelii siamensis*) has a direct nephrotoxic action that can manifest itself even in the absence of any systemic symptomatology. This nephrotoxicity must be distinguished from a complication of the haemorrhagic syndrome that often leads to a destruction of the renal tissue. Monovalent antivenins are produced in Thailand and China and a polyvalent one in India.

The venom of *D. palaestinae* is also known to contain a phospholipase A_2 that can provoke neurotoxic symptoms.

Echis (saw-scaled viper). This genus is widely distributed from East Africa to Sri Lanka and to Central Asia. It consists of species that are difficult to distinguish from each other. The most important ones of the eight presently recognized species are *Echis carinatus* (saw-scaled viper) from India to Iran and Uzbekistan which many authors would divide into four or five related species, *E. coloratus* (painted carpet viper, Burton's carpet viper) from the Arabian peninsula, *E. pyramidum* (carpet viber) from the Maghreb to Egypt and Djibouti, *E. leucogaster* (white-bellied carpet viper) in the African Sahel and the oases of the Sahara and

E. ocellatus (West African carpet viper) in the African savannah of the Sudan, where it is also a complex of species. In India and in Sri Lanka *E. carinatus* accounts for 80 to 95% of the bites, like *E. ocellatus* to which more than 85% of envenomations are attributed in sub-Saharan Africa. In Saudi Arabia *E. coloratus* causes 60% of the snake envenomations.

The venom of the *Echis* spp. contains proteolytic enzymes that cause inflammation problems and localized necroses, and a prothrombine activator which provokes a severe and prolonged haemorrhagic syndrome. Many polyvalent antivenins are produced in France for Africa, in Saudi Arabia and in Iran for the Near and Middle East and in India for Central and South Asia. Recent investigations have shown a certain cross reactivity in the antivenins prepared for the venoms of *Echis* spp. from far apart regions which translates into a cross efficacy. The main difference between the antivenins resides in the degree of their purification and consequently of tolerance.

Macrovipera. This genus was formerly part of the genus *Vipera*. It consists of two species distributed from Algeria, where only relict populations are observed, to Central Asia. The most widely spread one is *Macrovipera lebetina* (Levant viper).

The venom of the *Macrovipera* spp. is strongly inflammatory and necrotizing. No particular haematological problem has been described. A polyvalent antivenin is produced in France for the Maghreb. Monovalent antivenins are produced in Algeria and in Iran.

Montivipera. This newly validated genus was formerly part of the genus *Vipera* and consists of eight species, all from the Near and Middle East. The most important species are *Montivipera raddei* (Radde's mountain viper) in Armenia and *M. xanthina* (Ottoman viper, coastal viper) in Turkey.

Vipera. This genus is widely distributed in the Old World from Great Britain to Finland and Korea in the north to Morocco and Iran in the south. The main ones of the 24 described species are: *Vipera aspis* (asp viper) in Eastern Europe, *V. ammodytes* (sand viper) and *V. ursinii* (field adder) in Central Europe, *Vipera berus* (adder, northern viper) in Northern Europe, *Vipera latastei* (Lataste's viper) in Southern Europe and North Africa.

The venom causes strong inflammation, weak haemorrhages and is sometimes neurotoxic, notably in *V. ammodytes* and in some *V. aspis* populations. This neurotoxicity is caused by an phospholipase A_2 that can show a presynaptic toxicity of varying intensity depending on the individuals. Polyvalent antivenins are produced in France, Germany, Croatia and the United States.

3.2.4.3 Crotalinae
This subfamily looks like the Viperinae but is distinguished by the fact that the former have a loreal thermo-sensitive pit. They are divided into three characteristic tribes.

The tribe of the Crotalini is characterized by the presence of a caudal rattle. They are divided into two genera, *Crotalus* and *Sistrurus*, and occur in North America and to a lesser extent in Latin America.

Crotalus (rattlesnake). This American genus presents the traditional rattlesnake. Generally the head is covered in small scales except the supra-ocular shields that are large and sometimes the frontal ones which can be larger than the other scales. It has several species in the north and center of the continent, and amongst them *Crotalus horridus, C. adamanteus, C. scutulatus* and *C. viridis* are the most important ones. The venom contains many enzymes that cause very severe local inflammatory symptoms, necroses and a severe haemorrhagic syndrome. Also, several North American species (*C. atrox, C. horridus, C. scutulatus*) can sometimes cause a neurotoxicity syndrome. The symptomatology varies but often includes a more or less severe fibrillation (repeated, short contractions of low amplitude) of the skeletal muscles, sensory problems (pinpricks, tingling) and a deep tiredness.

C. durissus (cascabel) is the only species found in Latin America. Several subspecies have been described, notably *C. durissus durissus* in Central America with a venom which is very similar to that of the other *Crotalus* spp. and *C. durissus terrificus*, widely distributed from Venezuela to Argentina on the eastern side of the Andes which has a myotoxic and neurotoxic venom. This subspecies accounts for 10 to 40% of the bites depending on the locality. The envenomation is characterized by indistinct local symptoms, respiratory problems and a diffuse microscopic muscle lysis that can lead to renal insufficiency due to tubulo-necrosis. Polyvalent antivenins are produced in the United States, Mexico, Costa Rica and Brazil.

Sistrurus (pigmy rattlesnake). This strictly North American genus occurs from Canada to Mexico. It also characteristically has a rattle, but the head is covered by large shields that resemble those of the Colubridae. On average they are smaller than the *Crotalus* spp. and their venom is less toxic. The genus has three species, the main two of them being *Sistrurus catenatus* and *S. miliaris*.

The venom causes strong inflammation and weak necroses and haemorrhages. There is no specific antivenin but there is a good cross reactivity with the *Crotalus* and *Bothrops* antivenins.

The tribe of the Lachesini has no tail rattle; the head is always covered by small, keeled scales. This tribe consists of the genera *Lachesis, Bothrops sensu lato*, both endemic in Latin America, and *Trimeresurus sensu lato* found in Asia and split into four genera: *Ovophis, Protobothrops, Trimeresurus* and *Tropidolaemus*.

Bothrops (fer-de-lance, lancehead). This South American genus occurs widely throughout Latin America. It comprises 39 species, of which the principal ones are: *Bothrops asper* (terciopelo), *B. atrox* (common lancehead), *B. bilineatus* (green forest-pitviper), *B. brazili, B. jararaca* (jararaca), *B. jararacussu* (jararacussu), *B. lanceolatus* (Martinique lancehead), *B. moojeni, B. neuwiedi*. These species occur practically in all biotopes, from the primeval forest (arboreal *B. bilineatus* and terrestrial *B. brazili*), to human settlements including the suburbs of large cities like *B. asper, B. atrox, B. jararaca* and *B. lanceolatus*. The frequency of bites varies with the regions. In Brazil, Venezuela, Guyana and Ecuador *B. atrox* causes from 40% (Amazonia) to 80% (Minas Gerais) of the bites. In Central America *B. asper* causes 20 to 60% of the envenomations. In the State of Sao Paulo *B. jararaca* causes 95% of the envenomations.

The venom is rich in proteolytic and haemorrhage-causing enzymes, notably the thrombin-like enzymes that cause blood coagulation. The symptomatology includes strong inflammation, necrosis and haemorrhage. Many antivenins of varying quality, efficacy and tolerance are produced in Central and South America. The venom of *B. lanceolatus* causes diffuse micro-embolisms that can lead to visceral and notably cerebral infarcts. A specific antivenin is produced in France for this species which is endemic in Martinique.

Lachesis (bushmaster). This Latin American genus inhabits forest and dry land and consists of three species of which the two most important ones are *Lachesis muta* in South America and *L. stenophrys* in Central America.

The venom causes severe inflammation and sometimes necrosis. Bites appear to be rare, probably as these species remain relatively shy. However, in some regions of South America *L. muta* has been blamed for 5 to 30% of the bites. In spite of the imposing size of some individuals, the envenomations are less severe than one could fear and what would be suggested by their reputation. Polyvalent antivenins are produced in Mexico and Brazil.

Protobothrops (habu). This genus comprises eight species that are widely distributed from India to Japan. The main species are *Protobothrops flavoviridis* and *P. mucrosqamatus*. The former is responsible for 20% of the envenomations in Japan and the latter for 40 to 50% of the bites in Taiwan and Malaysia.

The venom causes severe inflammation, necrosis and haemorrhage. It contains a thrombin-like enzyme that accelerates coagulation and provokes haemorrhage after all the coagulation factors have been used up by the venom. A specific antivenin is produced in China against the venom of *P. mucrosquamatus*.

Trimeresurus (Asian pitviper, bamboo pitviper). This genus is distributed from Sri Lanka to Nepal and from Japan to Indonesia and contains 29 species, the most important of which are: *Trimeresurus albolabris, T. gramineus, T. popeiorum, T. purpureomaculatus, T. stejnegeri*. In Thailand *T. albolabris* is implicated in 40% of the bites while in Hong Kong it is blamed for 60% of the envenomations.

The venom contains powerful proteolytic enzymes and a thrombin-like enzyme. The symptomatology includes signs of severe local inflammation, sometimes complicated by gangrene or necrosis, and a haemorrhagic syndrome that can be reduced by symptomatic treatment, even without the help of an antivenin. A monovalent antivenin is produced in Thailand and a polyvalent one in China.

The tribe of the Agkistrodontini was formerly together within the genus *Agkistrodon sensu lato*; They are distributed in Asia (*Calloselasma, Deinagkistrodon, Gloydius* and *Hypnale*) and in North America (*Agkistrodon*).

Agkistrodon (mocassin). This North and Central American genus contains three species of which the two with the most widespread distribution are *Ackistrodon contortrix* and *A. piscivorus*. Several hundred bites occur each year in the southern United States. The venom causes inflammation and haemorrhage. It is moderately toxic; without specific treatment the lethality is below 0.1%. The polyvalent antivenins produced in France have a cross reactivity which is considered sufficient in cases of envenomation with severe symptoms.

Calloselasma (Malayan pit viper). This Far East genus is monospecific: *Calloselasma rhodostoma*. This diurnal species abounds in all of South East Asia and causes a large number of severe envenomations, particularly in rubber plantations. In Thailand it is blamed for 40% of the bites and in Indonesia up to 70% of envenomations are due to this species.

The venom is moderately toxic and causes severe inflammation, necrosis and haemorrhage. The first thrombin-like enzyme to be extracted from a snake venom came from this species. Several monovalent antivenins are produced in China and Thailand and polyvalent ones in India and Indonesia.

Deinagkistrodon (hundred-pace-viper). This Far Eastern genus has only one species: *Deinagkistrodon acutus*. Its common name comes from a legend according to which the victim could not go further than one hundred paces before collapsing.

The venom causes inflammation and haemorrhage. Lethality is high, more than 15% without antivenin treatment. A monovalent antivenin is produced in China.

Gloydius (mamushi). This genus is distributed from Afganistan to Japan and comprises nine species of which the most widespread ones are *Gloydius blomhoffi* and *G. halys*. In some regions of China and Japan *G. blomhoffi* is responsible for up to 80% of the bites.

The venom causes inflammation. Symptoms of neurotoxicity are sometimes observed. The main complication is renal insufficiency that sometimes is fatal. A monovalent antivenin is produced in Japan.

Hypnale. This genus occurs in Sri Lanka and contains species of small size. They cause a limited number of envenomations of moderate severity.

PART II
Venoms

CHAPTER 2

The Venom Apparatus

The venom apparatus is a complex device consisting of a specialized gland that synthesizes a toxic secretion, the venom, and a wounding device, the venom fang, which can inject the venom into the body of the prey or of the aggressor. This structure is particularly complex in snakes.

The venoms may have derived from a specialization of digestive secretions, possibly pancreatic, certainly salivary, which originally led to the digestion of the tissues. The saliva has a twofold role: it lubricates the food and it starts the process of digestion. Later the venoms would have developed the capability to kill and to immobilize the prey with the help of specialized toxins facilitating holding and swallowing, which are difficult without the help of legs. In any case one can think that the role of the venom in the snake's defense is secondary, even if it is the one that concerns us most.

1. The General Evolution of the Venom Apparatus

The evolution of the venom apparatus will on one hand concern the position and morphology of the fang and on the other hand the structure of the venom gland and the composition of the venom.

It is generally conceded that the venomous function has appeared relatively late and progressively starting from the differentiation of the salivary glands. From the point of view of morphology as well as of biochemistry it appears now well established that the venom apparatuses of the principal families of venomous snakes have evolved separately: proteodontism in the Elapidae and opisthodontism in the Colubridae, Atractaspididae and Viperidae. Also, this specialization appears to have been relatively ancient and on the way to disappear in certain groups, the evolution tending, according to some authors, to suppress the venomous function.

1.1. The evolution of the maxilla

In the aglyphous snakes which do not have poison fangs, the proterodont ones have teeth which decrease in length from front to back, while on the maxilla of the opisthodont ones the teeth increase in length from front to back. In both the apparition of a groove or canal along the length of one or several teeth allows the penetration of the saliva into the tissues of the prey. In different species this groove consists of an open canal (grooved fang) or becomes closed completely to form a tube (tubular fang), which allows the injection of the venom under pressure. In all cases the fang progressively becomes separated from the other maxillary teeth (Fig. 9). This more or less marked distance between the fang and the full teeth of the maxilla is called diastem. In the course of this evolution the total number of teeth is reduced, ultimately leaving only the fang.

Proterodontism assists the application of maxillo-prefrontal pressure, probably weaker but facilitating holding the prey, which means a certain advantage. This route leads to the proteroglyphous, front-fanged, snakes to which the Elapidae belong.

The opisthoglyphous route is first observed in the opisthoglyphous Colubridae. The maxilla remains elongated even though it is shortened in comparison to fangless opisthodont snakes. The fang is situated at the back of the maxilla, at the level of the eye or further back. The fang is grooved with normally a wide-open groove.

In a more advanced stage the opisthoglyphous route leads to the solenoglyphous venom apparatus of the Viperidae. The maxilla is considerably shortened and carries only the single fang with replacement fangs ready to take the place of the functional fang if it should be lost. The venom fang is tubular and can reach an impressive size. Because of that it has become mobile in the prefrontal articulation: at rest the fang lies horizontally and is erected at the moment of the bite.

1.2. The evolution of the venom gland

The advent of the salivary glands, be it in mammals or in reptiles, goes back 200 million years, which seems relatively recent. In the venomous snakes one probably has to look for the origin of the venom glands in a specialization of the labial glands. Iguanas as well as amphisbaenes have a series of glands both on the upper

The Venom Apparatus

Figure 9
The dentition of snakes

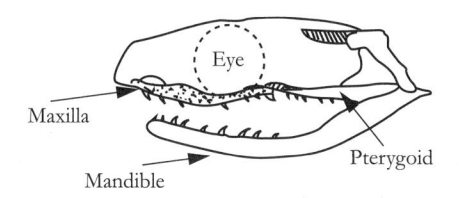

Aglyphous (fangless):
– full teeth,
– lack of venom gland,
– boas, pythons and most of the colubrids.

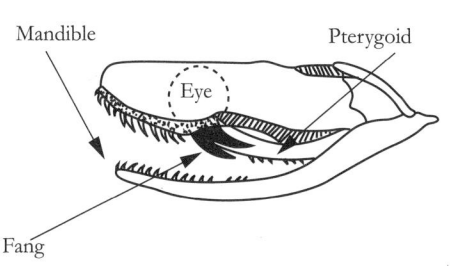

Opisthoglyphous (back-fanged):
– fangs back on the maxilla (at the level of the eye) and with a median groove,
– presence of venom glands,
– the remainder of the colubrids, haemotoxic venom.

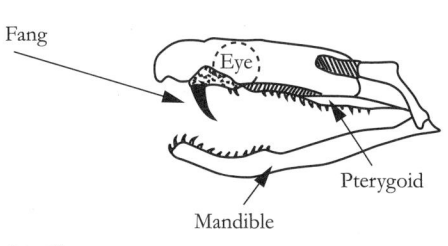

Proteroglyphous (front-fanged):
– fangs in the front part of the maxilla that is immobile,
– presence of venom glands,
– cobras and mambas, neurotoxic venom.

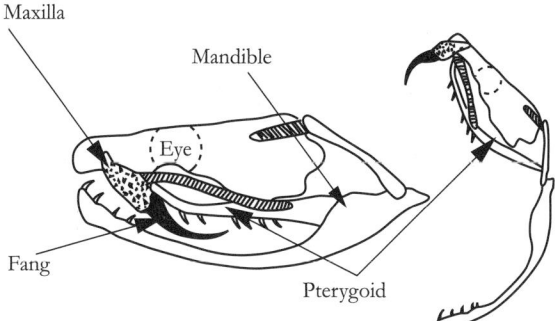

Solenoglyphous (hollow-fanged):
– fangs in the front part of the maxilla that is mobile,
– presence of venom glands,
– vipers and crotalids, haemotoxic and necrotising venom.

and lower lips. The evolution of the upper lip glands in the direction of more complex enzyme secreting glands would have taken place at the same time as the modification of the cranial bones, of the muscles and of the teeth. In addition to the flaccidity of the ligaments allowing a wide opening of the mouth common in the Alethinophidians, the development of the maxilla and of the muscles has led in the venomous species to the establishment of a sophisticated mechanism for the injection of the venom and of specific muscles for the compression of the venom glands. Certain glands open directly at the base of the teeth and are called dental glands. They can be considered as precursors of the complex glands observed in snakes and in those lizards that have developed a venom function like the heloderms. Some authors believe in any case that these glands cannot be likened to the proper venom glands, which they rather see as derived from the Duvernoy's gland. This latter which is very ancient and also occurring in other groups than snakes, notably in certain lizards, does not necessarily correspond to a venom gland as its role is neither clear-cut nor constant. The Duvernoy's gland produces a liquid which is between saliva and venom, the purpose of which could be limited to the lubrication of the prey or to its digestion but also to its immobilization and to killing it quickly, even more so as in this latter case it proves to be of unequal efficacy. The development of specialized glands is concomitant with the evolution of the venom itself. The first active substances have probably been enzymes with the essential aim to digest the prey.

The whole of the histological structures could also have come from the evolution of another exocrine gland like the pancreas. Many of its enzymes such as amylases, phospholipases or proteases occur also in the salivary glands and in the venom glands. There is a variable degree of homology, but it can reach 90% for certain enzymes. It appears that some have evolved secondarily to be transformed into toxins. The cytotoxins of the Elapidae, for example, could have come from the Phospholipases A2. For other authors the neurotoxins could have been precursors of the cytotoxins. The origin of the sarafotoxins present in the venoms of the Atractaspididae would be more complex: they would have derived from endothelins, which are considered as mediators at cellular level acting on the blood vessels and permitting vasoconstriction.

The development of toxins from pancreatic enzymes also helps to explain the presence of enzyme inhibitors in the blood that protect the snake against its own venom when it escapes from the venom gland and penetrates into the blood circulation. These proteins found in the serum of snakes and in charge of neutralizing the venom, generally are albumins or α-globulins with a high molecular weight. One can thus imagine that in snakes there could have been a process of co-evolution of on one hand the toxins derived from digestive enzymes and on the other hand the antitoxins deriving from enzyme inhibitors. An analogous situation is also observed in certain mammals that possess a natural resistance to snakebite venoms like mongooses, marsupials and hedgehogs.

2. The Anatomy of the Venom Apparatus

The scolecophidians are considered as the most primitive living snakes. They have no toxic secretion in their oral cavity. Apparently they also do not have dental glands and could possible represent an evolutionary line in which these glands have disappeared.

The Boidae are constrictors; they kill their prey by suffocation. Often their saliva is toxic, but they lack an inoculating mechanism. Also they are characteristic examples of the precursors of the proterodont route.

The Colubridae have a large variety of diverse genera and species. The differentiation of the labial glands probably has preceded the specialization of the teeth. All the same the fangless species of Colubridae have highly toxic venomous secretions produced by their Duvernoy's glands. However, there are snakes with differentiated teeth but without a Duvernoy's gland. In the Opisthoglypha the groove is situated in the prolongation of the efferent canal of the Duvernoy's gland. However, the injection mechanism is quite different from the one common to other venomous snakes. On one hand the Duvernoy's gland does not have a wide lumen, which reduces its capacity to store venom, and on the other hand there is no muscle directly associated with the gland. Consequently the injection mechanism is not very efficient and it is rather by gravity and capillarity that the venom penetrates into the bite wound while the prey is being held and swallowed.

In the Colubridae the Duvernoy's gland varies from species to species in size, volume and histological structure. The largest and probably the most efficient one is seen in *Dispholidus typus*. The same is the case in the genus *Thelotornis* that also has enormous posterior fangs with a deep groove. In addition the venom of these two species is particularly toxic. The Duvernoy's gland is a tubular, branched gland with serous secretions produced in the tubules and mucous secretions consisting of mucopolysaccharides appearing in the epithelial cells of the excretory canals. It lacks a storage reservoir. The secretions of the Duvernoy's gland have multiple enzymatic activities. One finds proteases, a cholinesterase, phosphatases, a hyaluronidase, a collagenase, phospholipases and various enzymes acting on blood coagulation. In some species of opisthoglyphous colubrids the venom appears to be particularly toxic, rapidly immobilizing the prey, which makes it easier to hold it. Nevertheless, in the majority of the colubrids the venom acts very slowly and kills the prey in 24 or 48 hours, as in the case of *Malpolon*, the Montpellier snake. It is possibly also simply the quantity of injected venom that produces such differences.

The Elapidae are a family of venomous snakes in the strict sense. The skull is similar to that of the Colubridae, with notably the pterygoid of the same size. In most of the proteroglyphs the maxilla is short and has one or several poison fangs with almost closed grooves or a closed tube, followed by 1 to 15 full teeth behind a diastem of variable size. In *Dendroaspis* as well as in certain Australian elapids the maxilla has remained relatively long and has acquired a certain mobility that allows it a certain see-saw movement around the pterygoid.

The structure of the venom gland of the Elapidae shows characteristics that are considered as primitive. The venom gland has differentiated itself from the upper labial gland. Surrounded by a fibrous sleeve it has migrated to behind the eye in the temporal region where its pear shape is seen prominently on either side of the head. It is a lobulated gland composed of serous cells arranged around acini. The central lumen is relatively small and the secretions are stored in cytoplasmic granules before reaching the lumen of the gland the volume of which increases. In this the histological structure of the secretory cells is very similar to that of the Duvernoy's gland. Also the accessory mucous glands are situated along the efferent tract and flow into the excretory canal. The latter opens at the base of the fang. The expulsion of the venom is effected by the contraction of a special muscle which derives from the mandibular and temporal muscles. The principal difference with the Colubridae consists in the composition of the venom. This confirms that the glandular epithelium retains the genetic information needed for the production of the glandular secretions. The venom of the Elapidae is rich in toxins acting on specific cellular receptors and contains a relatively small number of enzymes. Some of those, amongst others certain phospholipases, have a toxic activity and at the same time their own enzymatic activity.

In the Viperidae the venom apparatus is very different, in its structure as well as in its functioning, which probably points at a distinct phylgenetic origin. This would confirm the hypothesis that the Viperidae and the Elapidae descended from separate evolutionary lines. Other arguments tend to associate the venom apparatus and the venom of the Colubridae with those of the Viperidae which could have descended from a common ancestor. The particularly important looseness of the bones of the skull in the Viperidae allows them to hit and to swallow prey animals of very large size. The maxilla is short; it pivots around the ectopterygoid supported by the pterygoid, two jaw bones of snakes, and this allows the poison fang to be turned forward. The latter is very long and forms a completely closed tube that opens at the point of the tooth. A sleeve of mucosa surrounds the tooth and allows the opening of the canal from the poison gland to connect with the tube at the base of the fang. The strong muscles surrounding the poison gland assure an injection under pressure. In addition the length of the fang enables a deep penetration.

The venom gland is situated in the temporal region. It is well visible on both sides of the head. It is a lobulated serous gland. It has been shown in *Echis pyramidum* that each of the cells of the central portion of the gland is able to synthesize all the constituents of the venom. These are produced successively by the secretory cell and/or at different speeds, which can explain a variation of the toxicity of the venom during the secretory cycle. The lumen of each lobe allows the storage of a large quantity of venom. The excretory canal has bulging portions where mucus is secreted. It opens directly at the base of the poison fang. The venom gland is surrounded by its own muscles that have derived from the temporal muscles. The particularly fast toxic action of the venom completes the device designed to prevent the prey from escaping. The venom has a complex composition and consists principally of many and varied enzymes.

The venom apparatus of the Atractaspididae is similar to that of the Viperidae. However, it distinguishes itself probably because of the burrowing habits of these snakes. The *Atractaspis* hit their prey laterally down and backwards with a single fang that passes through the discretely opened corner of the mouth. Also the pterygoid is short and separated from the palate, which allows it to increase the rotational movement of the maxilla during the erection of the fang.

The venom gland of the Atractaspididae differs from that of the Elapidae and the Viperidae. The secretory tubules are arranged around an elongated and narrow lumen and the mucus cells are at the exit of the gland. There is no accessory gland and several species have extremely long glands that can stretch one third of the length of the snake along neck and thorax. The venom contains enzymes and one particular type of toxins: sarafotoxins.

3. The Composition of the Venom

Snake venom is a complex mixture of proteins. Their fine analysis has been achieved progressively only as permitted by technological progress.

It is obtained by manual pressure on or electrical stimulation of the salivary glands of the snakes. Recently cell culture of the glandular cells has made possible the artificial production of venom in small quantities. However, certain protein fractions can also be synthesized by recombination in molecular biology at acceptable cost. To this end microorganisms are used into which a gene from the venom gland is inserted which codes for a particular protein.

Native venom is dried in a vacuum or lyophilized, and can be conserved in a sealed ampoule at low temperature for several decades, although certain enzyme activities appear to change over time irrespective of the process of conservation.

3.1 The synthesis of the venom

The production of venom passes through a first phase of rapid synthesis and after that progressively reaches a plateau that seems to correspond to the saturation of the gland. Each secretory cell produces all the constituents of the venom. The maximum of venom synthesis is reached more or les within a week and the plateau in two or three weeks. After a period of strong production of the venom constituents the re-absorption of water allows the protein concentration to be stabilized. The synthesis of the venom stops when the central lumen of the gland, which acts as reservoir, is full.

It is normally not possible to gauge the quantity of venom injected into a prey, particularly in man. On one hand it varies with the species of snake in question (Table II) and on the other hand the circumstances of the bite appear have a strong influence on the quantity of venom that is injected. The proportion of injected venom during a bite can vary from 10 to 50% of the gland's capacity. It depends also on the size of the prey, its species and on the hunger of the snake. It has been shown that in certain species of *Crotalus* the amount of venom injected in a defensive bite is twice or three times of that injected when biting a prey. With successive bites the quantity of venom delivered by the snake diminishes gradually. All happens as if the snake delivered the necessary dose to obtain the desired result: immobilize the prey or eliminate an aggressor if a confrontation cannot be avoided.

The spray of venom by a *Naja nigricollis* delivers approximately 10% of the volume of the venom gland, which allows a very large number of consecutive sprays (up to 40 times in less than 5 minutes). In the first droplets one observes a protein of 9 kDa, which is never collected later, which is present in naturally collected venom (without utilizing any apparatus nor pressure on the glands) and found only if the snake has been excited before spitting the venom. The analysis of the proteins present in the venom droplets shows that the concentration of proteins decreases significantly with time. However, the concentration of cytotoxins remains more stable although it also decreases. It is as if there was a dilution of the venom but with the level of cytotoxin remaining almost constant.

3.2 The chemical composition of the venom

Traditionally proteins isolated from snake venoms are divided in two groups: the enzymes which generally are of weak acute toxicity and the toxins of which the pharmacological role is better known and which at least in the Elapidae and the Atractaspididae form the essence of the toxicity of snake venoms.

Table II
The capacity of the glands and the toxicity of the venom of the main venomous species

Espèce	mean glandulaire capacity (mg of dry venom)	LD50 intravenous (μg/mouse)
Agkistrodon contortrix	40 to 75	218
Agkistrodon piscivorus	80 to 170	83
Atractaspis enggadensis	25 to 75	5
Atractaspis dahomeyensis	25 to 75	45
Bitis arietans	100 to 150	7
Bitis gabonica	200 to 1 000	14
Bothrops atrox	80 to 200	91
Bothrops lanceolatus	100 to 200	300
Bungarus caeruleus	10 to 20	2
Causus maculatus	50 to 100	185
Cerastes cerastes	20 to 30	10
Crotalus adamanteus	200 to 850	34
Crotalus atrox	150 to 600	44
Crotalus durissus terrificus	150 to 500	84
Crotalus scutulatus	50 to 150	5
Daboia russelii	130 to 250	16
Deinagkistrodon rhodostoma	40 to 75	95
Dendroaspis polylepis	100 to 200	5
Dendroaspis viridis	50 to 100	12
Dispholidus typus	5 to 10	2
Echis carinatus	10 to 50	25
Echis leucogaster	20 to 75	35
Echis ocellatus	10 to 50	33
Lachesis muta	200 to 1 200	120
Micrurus fulvius	2 to 10	6
Naja haje	150 to 600	15
Naja kaouthia	140 to 400	8
Naja melanoleuca	150 to 800	12
Naja nigricollis	150 to 750	25
Notechis scutatus	30 to 70	1
Rhabdophis subminiatus	5 to 10	25
Sistrurus miliarius	10 to 35	57
Thelotornis capensis	1 to 5	15
Vipera aspis	5 to 15	20

The Venom Apparatus

3.2.1 Enzymes

The enzymes are proteins with generally elevated molecular weights. Their catalytic properties that distinguish them from the toxins have two major consequences. On one hand the degradation product on which in most cases the toxicity depends, generally has no immunogenic property in the receiving organism. Therefore it does not lead to the synthesis of specific antibodies. On the other hand the toxicological effects depend more on the time during which the enzymatic reaction is taking place than on the initial quantity of enzymes. They in fact cause a biological transformation without themselves being modified, which allows them to carry on reacting as long as they are present in the organism. This amplification is characteristic of the enzyme reaction and thus its toxic effects are essentially time-dependent. The enzymes of snake venoms vary in their specificity. The most toxic ones act on blood coagulation or on complement activation, provoke a cytolysis or accelerate a particular metabolism (phospholipids, glycides).

The venoms of the Viperidae are particularly rich in enzymes.

3.2.1.1 Phospholipases

The majority of snake venoms contain phospholipases that hydrolyze free phospholipids and those bound to membranes, into fatty acids and lysophospholipids (Fig. 10).

Several types of phospholipases are distinguished according to the site of the hydrolysis. In snake venoms the Phospholipases A2 are the ones that occur most

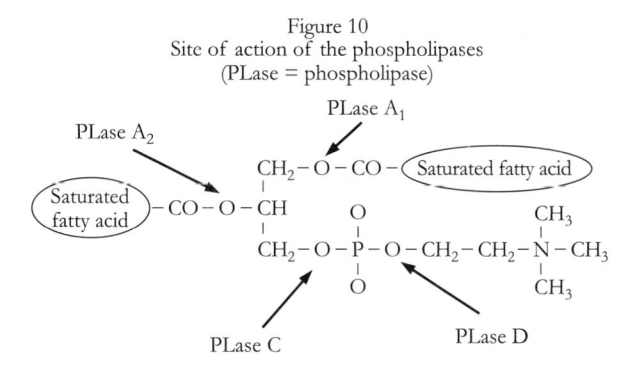

Figure 10
Site of action of the phospholipases
(PLase = phospholipase)

often. The lysophospholipids are tension-active and cause cellular destruction, particularly haemolysis. The Phospholipases A2 affect several physiological systems depending on the type of hydrolyzed phospholipids, haemostasis, neuromuscular transmission and inflammatory reaction.

The molecular weight of the base monomer is close to 8 kDa, but a polymerization of the molecule can raise the molecular weight to 36 kDa.

3.2.1.2 Acetylcholinesterase

The Elapidae have an acetylcholinesterase that can hydrolyze acetylcholin, which is the principal chemical mediator of the transmission of nerve stimuli in vertebrates (Fig. 11).

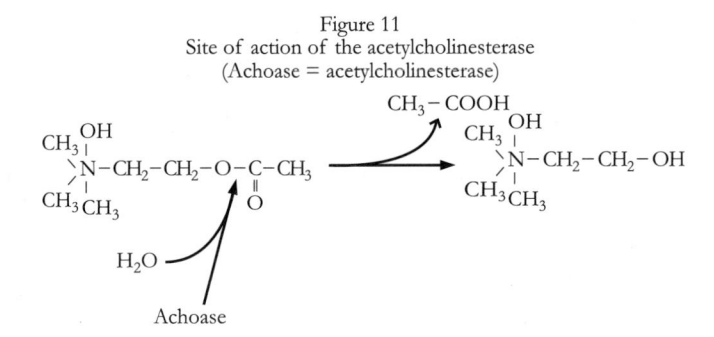

Figure 11
Site of action of the acetylcholinesterase
(Achoase = acetylcholinesterase)

This enzyme has a molecular weight of 126 kDa and consists of two monomers of 63 kDa with a di-sulfur bridge; it is active in an alkaline environment with a pH of 8 to 8.5. It plays an essential role at the synapse where it allows the passage of nervous stimuli to the postsynaptic membrane. It contributes to the complex neurotoxic effect of elapid venoms.

3.2.1.3 Phosphoesterases

Many venoms contain various phosphoesterases. The endonucleases hydrolyze the nucleic acids (DNA and RNA) at the bonds between the base pairs. The exonucleases attack the base at the end of the nucleic chain. They are active in an alkaline pH close to 9 or 10.

The Venom Apparatus

The phosphodiesterases cut the link removing the oxygen from position 3' of the ribose or the desoxyribose, separating them from the phosphorus. The 5'nucleotidase produces a similar cut but at the level of the bond 5' between the ribose or the desoxyribose and the phosphor (Fig. 12).

Finally the phosphomonesterases are less specific and hydrolyze all the mononucleotides, particularly those that are involved in energy transport in the cells.

Figure 12
Site of action of the phosphoesterases

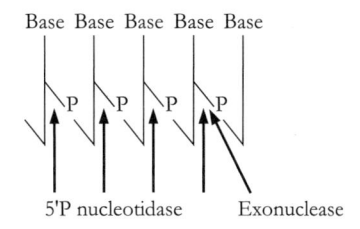

Base Base Base Base Base
— P — P — P — P —
5'P nucleotidase   Exonuclease

3.2.1.4 L-amino-acid-oxidase

This enzyme causes the des-amination and then the oxidation of the amino acids, which are transformed into α-cetonic acid (Fig. 13).

Figure 13
Site of action of the L-amino-acid-oxidase

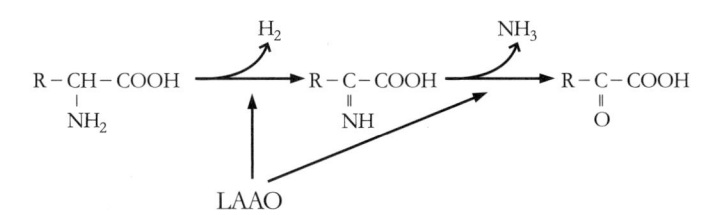

$$R-CH(NH_2)-COOH \xrightarrow{H_2} R-C(NH)-COOH \xrightarrow{NH_3} R-C(O)-COOH$$

LAAO

Its molecular weight is between 85 and 153 kDa. The clinical and toxicological effects are negligible; it represents less than 1% of the total toxicity of the venom, which is explained by the low concentration of this enzyme. The attached flavin-adenin-dinucleotide group of this enzyme gives the venom its yellow colour.

3.2.1.5 Hyaluronidase

This enzyme is common in most venoms. It hydrolyzes hyaluronic acid or the sulfate of chondroitin, which are mucopolysaccharides responsible for the cohesion of the connective tissue. Therefore this enzyme probably enhances the diffusion of the venom after its injection by a bite.

3.2.1.6 Proteases

There are many enzymes that act on the structure of proteins. The venoms of the Viperidae have particularly many of them. They are also involved in the tissue destruction seen during necrosis and in the course of certain pharmaco-toxicological phenomena like haemostasis problems, as shall be seen later. While it is true that many proteases are non-specific and act on a large number of different substrates, there are some of these enzymes that recognize particular molecular sites, which makes them particularly efficient tools for the diagnosis or treatment of certain conditions.

The proteases that act on blood coagulation are classed in two structural groups: the serin-proteases and the metallo-proteases.

- The serin-proteases are single- or double-chained glycoproteines of 20 to 100 kDa that attach themselves to their substrate with the help of a serin. They exist in great diversity depending on their zoological origins and their biological functions. Most of the thrombin-like enzymes belong to this group; generally their structure is similar to that of the chain B of natural thrombin. Certain fibrinolytic enzymes and the fibrinogenolytic enzymes that degrade the chain Bβ of fibrinogen also belong to this group. Finally one finds there certain enzymes that activate prothrombin or protein C.

- The metallo-proteases are single- or double-chained proteins of 20 to 100 kDa. The presence of a metal ion, generally zinc, is indispensable for their functioning. In addition calcium assures their structural stability. These enzymes are very sensitive to pH variations and are inactivated in an acidic environment (pH < 3). There are four groups of metallo-proteases. The first ones are composed of proteins that do not act on blood coagulation and are considered as "pro-peptides", precursors of the other metallo-proteases. The second ones, or the proper metallo-proteases, have a low molecular weight and haemorrhage causing properties. The 3rd group consists of metallo-proteases of the 2nd group with an additional peptide chain with a composition and structure identical to the disintegrins. Finally into the last group are placed complex metallo-proteases, which in addition to those of the 3rd group have a domain rich in cystein which accentuates the stability of the molecule.

The haemorrhage-causing activity and the specificity of the metallo-proteases are in proportion to their molecular weight. Several fibrinolytic enzymes and the fibrinogenolytic enzymes that degrade the Aα chain of fibrinogen are metallo-proteases with a molecular weight varying between 20 and 70 kDa. The metallo-proteases with a molecular weight between 50 and 70 kDa activate factor X or prothrombin. They exert their principal catalytic activity on the vascular endothelium and because of this some have been given the name haemorrhagins.

3.2.1.7 Other lytic enzymes
Amylase, transaminases and deshydrogenases are also found in small quantities, but all of these have only a negligible pharmacological and toxic effect in human pathology.

3.2.2 Toxins
Toxins are proteins of variable molecular weight, but generally less than 30 kDa, namely smaller than the enzymes. They have the ability to bind to a specific receptor, most often a membrane. The tropism of the toxins can be neurological, cardiovascular, muscular or undifferentiated depending on the anatomical distribution of the receptors recognized by them. The toxicological effect depends on the proportion of the introduced quantity of toxin to that of the corresponding receptors: it is considered as dose-dependant. Other factors can play a role, particularly the speed of diffusion of the toxin, which again depends strongly on its size and its specificity for the receptor. It should be noted that the quantity and the specificity of the receptor can vary from one animal species to another and that consequently the effect varies with the experimental model. The resultant of these factors leads to what one can call a target effect which establishes for a toxin in a given model a linear relation between the quantity of toxin, the number of available receptors and the pharmacological effects, namely the toxicity.

The venoms of the Elapidae are particularly rich in toxins.

The toxins can be placed into eight principal families depending on their structure and/or their mode of action. The neurotoxins are a particularly well-studied class of proteins (Table III). Many toxins belonging to different functional groups show a "3-finger" configuration or one similar to that called "pear-shaped" or "tangled".

Table III
The classification of the neurotoxins of snake venoms

Espèces	Ntx α- short	Ntx α-long	Ntx β	Dendrotoxin	Fasciculin	Cardiotoxin
Acanthophis antarcticus	toxin C					
Bungarus caeruleus		toxin b				
B. fasciatus			ceruleotoxin			
B. multicinctus		α-bungarotoxin	β-bungarotoxin			
Boulengerina annulata				homologous		
Dendroaspis angusticeps				homologous		
				DTx, $C_{13}S_1C_3$	FAS 1, 2, F7, $C_{10}S_2C_2$ S_6C_4	
D. jamesoni	$Vn^1 I$	V^nIII-1				
D. polylepis	toxin-α	Vn^1N, V^nI, V^n2		toxin I, K, B	toxin C, FS 2	
D. viridis	tox. 4.11.3	tox 4.9.3, 1, V		DV14	toxin 4.9.6.	
Enhydrina schistosa	toxin 4, 5		Myotoxin			
Hemachatus haemachatus	II and IV		unnamed			
Hydrophis ornatus	toxin 73 A			homologous		
H. coggeri						
H. cyanocinctus	hydrophitoxin					
H. gracilis						
Lapemis curtus						
Laticauda colubrina		toxin a, b				
L. laticauda	laticotoxin					
L. semifasciata	erabutoxin	LS III				
Micrurus corallinus			toxin I			
M. fulvius						
M. hemprichii						
M. lemniscatus						
M. narducii						
M. surinamensis						

Species					
M. nigrocinctus					
Naja haje	CM-10A, 14, α-toxin	CM-5, III		CM-10, 12	CN-7, 8, 9, 10B, V^{II}1, 2, 2A, CM
N. kaouthia	toxin-d	3 α-cobratoxins			CM-6, 7, 7A V^{II}1, 1A
N. melanoleuca	tox. I and III	toxin b, 3.9.4	CM II, III		V^{II}1, 2, 3, 4
N. mossambica	cobrotoxin		DE I, II, III		CM-XI, cobramin A, B
N. naja		toxin 3, 4, A, B, C, D, E	CM I, II, III		cardiotoxin A, B
N. nigricollis	α-toxin		unnamed		V^{II}1
N. nivea	toxin δ, β	unnamed			unnamed
N. oxiana	toxin II	α-toxin			
N. sputatrix			homologous		
Notechis scutatus		III-4	notexin		
Ophiophagus hannah		toxin a, b			
Oxyuranus scutellatus			taipoxin, OS 2		
Pseudechis australis					
P. butleri					
P. papuanus					
P. porphyriacus			pseudexin		
Pelamis platurus	pelamitoxin				
Pseudonaja textilis			textilotoxin		
Bitis caudalis			caudoxin		
Crotalus terrificus			crotoxin		
Daboia russelii siamensis			daboiatoxin		
D. russelii formosensis			RV-4, RV-7		
Gloydius blomhoffi			agkistrodotoxin		
Protobothrops mucrosquamatus			trimucrotoxin		
Vipera ammodytes			ammodytoxin, vipoxin		
V. aspis			unnamed		
V. palestinae			unnamed		

These proteins of 57 to 74 amino acids and 3 to 5 di-sulfur bridges are a particularity of the venoms of the Elapidae and are related to each other and the most important one amongst them is the α-neurotoxin.

3.2.2.1 Postsynaptic neurotoxins

The curare-like toxins act on the postsynaptic membrane and block the transmission of nervous stimuli by attaching themselves directly to the acetylcholin receptor. They consist of a single polypeptide chain of 60 to 74 amino acids and have a molecular weight of about 7 to 8 kDa, and are folded in 3 loops that are held in place by di-sulfur bridges (Fig. 14).

Three types are recognized: the short α-neurotoxins (group I: 60-62 amino acids, 4 di-sulfur bridges), the long α-neurotoxins (group II: 66 to 74 amino acids, 5 di-sulfur bridges) and the k-neurotoxins (66-70 amino acids, 5 di-sulfur bridges). The long α-neurotoxins and the k-neurotoxins show much structural homology but differ on one hand by their composition of amino acids and on the other hand by the former ones being monomers and the latter ones dimers. This gives them slightly different pharmacological properties, particularly regarding the recognition of the cholinergic receptor. While the α-neurotoxins are largely found in most of the Elapidae, the α-neurotoxins are present only in the *Bungarus* spp.

The muscarin-like toxins are a dozen small polypeptides isolated from the venom of Dendroaspis (mamba). Their structure is similar to that of the curare-like neurotoxins. They have a molecular weight of 7.5 kDa (approximately 65 amino acids, 4 di-sulfur bridges) and their name derives from their strong affinity to that part of the cholinergic receptor.

Finally some "unclassifiable" toxins have been isolated, notably from venoms of the Viperidae. Vipoxin, isolated from the venom of *Daboia russelii*, has a molecular weight of about 13 kDa. Its molecular structure is unknown. It does not act on the nicotine or muscarine receptors, but on the adrenergic receptors, which is exceptional for snake venoms.

3.2.2.2 Cytotoxins

Approximately 50 polypeptides form this group that is relatively homogenous because of their greatly homologous structure. The cytotoxins, also called car-

The Venom Apparatus 65

Figure 14
Molecular structure of an α-neurotoxin

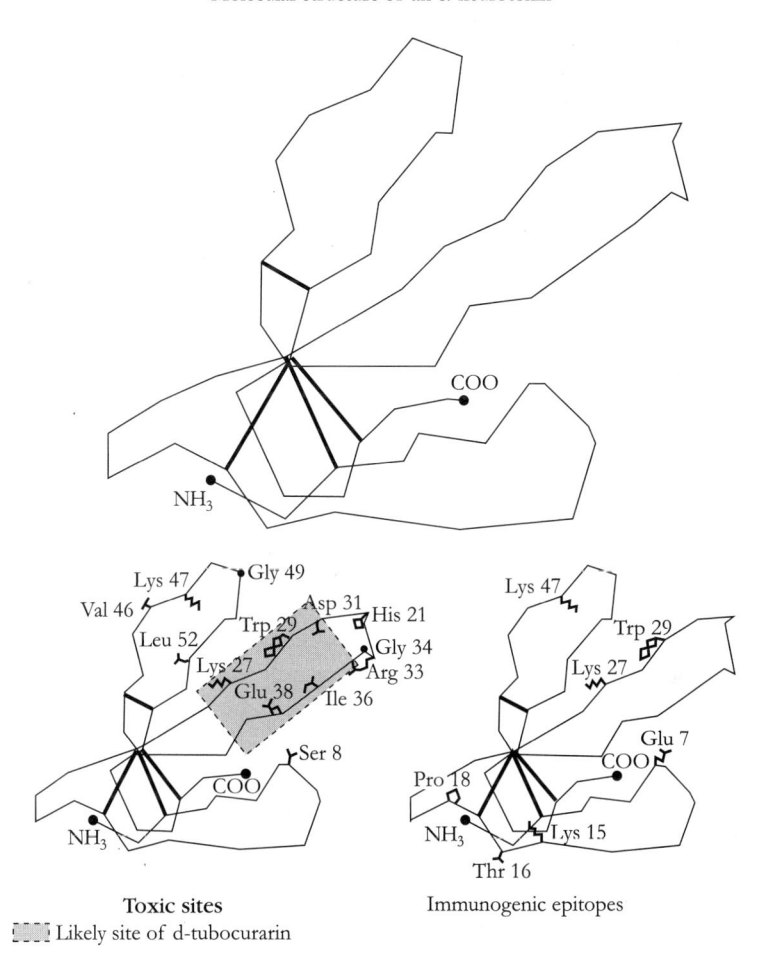

Toxic sites
░░░ Likely site of d-tubocurarin

Immunogenic epitopes

diotoxins, are in structure and size similar to the short α-neurotoxins (60-65 amino acids, 4 di-sulfur bridges, same shape in 3 loops). However, there is no immunological cross reaction with the neurotoxins. They depolarize rapidly and irreversibly the cell membrane causing its lysis.

3.2.2.3 Pre-synaptic neurotoxins

The pre-synaptic blocking toxins or β-neurotoxins consist of one, two or four sub-units, depending on their specific origin. They all have in common a Phospholipase A2 function needed for their toxic activity. The elementary subunit consists of a polypeptide chain of 120 amino acids with 6 to 8 di-sulfur bridges and with a molecular weight of 13 to 14 kDa. The β-neurotoxins are classified in three groups.

- **The single-chain β-neurotoxins** have only one phospholipase subunit. Some show an amino acid sequence similar to that of the Phospholipases A2 found in the mammalian pancreas. They are represented by notexin isolated from *Notechis scutatus*, an Australian elapid, as well as by the toxins extracted from the venom of several other Australian elapids (notably *Oxyuranus scutellatus* and *Pseudechis porphyriacus*). The others consist of a polypeptide chain of the same length but with an amino acid sequence that differs from that of the pancreatic phospholipase. Essentially these toxins are extraxted from the venoms of Viperidae: *Gloydius blomhoffi*, *Bitis caudalis*, *Vipera ammodytes* and *Daboia russelii siamensis*.

- **The β-bungarotoxin** from the venom of *Bungarus multicinctus* is formed by two different polypeptides linked by a di-sulfur bridge. One of the sub-units has a phospholipase activity; its amino acid sequence is similar to that of pancreatic phospholipase and has a molecular weight of 13 kDa. The second subunit with a molecular weight of 7 kDa is homologous with the dendrotoxin present in mamba venom.

- **The other known β-neurotoxins** are also composed of sub-units consisting of polypeptides. These polypeptides are identical or distinct, but their link is not covalent and none of the sub-units is homologous with the dendrotoxins. One of the polypeptides is an elementary polypeptide and has a phospholipase activity. A dozen of these toxins have been described which occur in the Elapidae as well as in the Viperidae (cf. Table III). The ceruleotoxin isolated from the venom of *Crotalus durissus terrificus* consists of two sub-units of different molecular structure, one of which has a phospholipase activity. The second sub-unit is smaller and lacks any toxic activity. It is associated with three peptides that separate themselves from the sub-unit when it binds to its membrane receptor. Many other American rattlesnakes have β-toxins with an analogous structure and the same could be the case with the venom of certain Old World vipers. Finally certain β-toxins such as taipoxin and textilotoxin isolated respectively from the venoms of *Oxyuranus scutellatus* and *Pseudonaja textilis* show a more complex structure. Taipoxine has three homologous sub-units and textilotoxin has four sub-units.

3.2.2.4 Dendrotoxins
The dendrotoxins, also called facilitating pre-synaptic toxins, were discovered in the venoms of *Dendroaspis* (mamba, an African elapid). They are proteins of low molecular weight (6 to 7 kDa) and consist of 57 to 65 amino acids with 3 di-sulfur bridges, which gives them a tangled structure. These toxins enhance the release of acetylcholine by blocking the voltage dependent potassium channels.

A structurally similar toxin, calcicludin, has been isolated from mamba venom. It also consists of about 60 amino acids and 3 di-sulfur bridges but its toxicological action is very different.

3.2.2.5 Fasciculins
These toxins have also been isolated from the venom of *Dendroaspis*. They are polypeptides of approximately 7 kDa molecular weight, with 60 or 61 amino acids and 4 di-sulfur bridges. Their structure is very similar to that of the α-toxins and the cytotoxins. However, none of the amino acids that constitute the active site of the curare-like neurotoxins is found in the same position in the fasciculins. There is also no immunological cross reaction with the antibodies directed against the α-toxins. The fasciculins are cholinesterase inhibitors and prevent the physiological regulation of nerve stimulus transmission.

3.2.2.6 Calciseptin
This toxin was isolated from the venom of *Dendroaspis polylepis* and consists of 60 amino acids with 4 di-sulfur bridges. The molecular structure is the same as that of the α-neurotoxins.

3.2.2.7 Myotoxins
Myotoxins are peptides of 42-45 amino acids (molecular weight 5 kDa, 3 di-sulfur bridges) without enzymatic activity. Essentially they are encountered in the venoms of *Crotalus* spp. They bind to the ion channels of muscle cells and cause

their necrosis. Crotamine extracted from *Crotalus durissus terrificus* acts on the sodium channel, while the myotoxin of *Crotalus viridis* affects the calcium channel. The myotoxin of *Philodryas* has a molecular weight of 20 kDa and consists of 182 amino acids with 3 di-sulfur bonds.

3.2.2.8 Sarafotoxins
The sarafotoxins have been isolated from the venom of *Atractaspis*, a Near Eastern burrowing snake. They are peptides consisting of about 20 amino acids and 2 di-sulfur bridges and a molecular weight of approximately 2.5 kDa. They are powerful vasodilators and are structurally and functionally very close to the endothelins, vasoconstrictor hormones present in the endothelial cells of mammals.

3.2.2.9 Disintegrins
The disintegrins found in the venoms of Viperidae are peptides of 5 to 15 kDa consisting of 49 to 84 amino acids with 4 to 8 di-sulfur bridges. They exist in mono- or heterodimeric conformation and according to their size are classified into three groups. The short disintegrins, like echistatin and eristostatin, extracted respectively from the venoms of *Echis carinatus* and *eristocophis macmahoni*, have 49 amino acids and 4 di-sulur bridges. The intermediary disintegrins of the type of tigramin from the venom of *Trimeresurus gramineus*, of kistrin from the venom of *Calloselasma rhodostoma*, of flavoviridin from *Protobothrops flavoviridis* and of botroxostatin from *Bothrops atrox* have 68 to 72 amino acids and 6 di-sulfur bridges. The long disintegrins, e.g. bitistatin isolated from the venom of *Bitis arietans*, are the most voluminous ones with 84 amino acids and 7 di-sulfur bridges.

These toxins inhibit the integrins, the trans-membrane proteins that allow the transfer of extracellular messages to the cytoplasm.

3.2.3 Other compounds

3.2.3.1 Nerve growth factor
The nerve growth factor (NGF) was serendipitously discovered in cobra venom and is a protein of 116 amino acids rolled up with 3 di-sulfur bridges. Frequently it occurs in dimeric form. It is essentially present in the venom of the Elapidae and has a molecular weight of 20 to 40 kDa. It has no enzyme activity and no toxicity. The nerve growth factor assists the differentiation of the sensorial neurons in the sympathetic ganglions.

3.2.3.2 Proteins which act on thrombocytes

Many molecules of varying molecular weight and structure act on the blood platelets or thrombocytes. They mask the effector sites on the cytoplasmic membrane or stimulate the degranulation of the thrombocyte.

The agregoserpentins are single- or double-chained glycoproteins of 25 to 80 kDa and a typical example is a protein isolated from the venom of *Trimeresurus gramineus*. The base subunit is a glycosilated polypeptide chain without enzymatic activity. They are equally present in the venoms of *Crotalus durissus terrificus* (convulxin), *Protobothrops mucrosquamatus*, *Tropidolaemus wagleri* (triwaglerin), *Bothrops jararaca* (botrocretin and bothrojaracin), *Calloselasma rhodostoma* (agretin) and *Trimeresurus albolabris* (alboagretin).

The thrombolectins have a molecular weight between 26 and 30 kDa. They consist of two identical subunits. Several thrombolectins have been extracted from the venoms of *Bothrops atrox*, *Crotalus atrox*, *Deinagkistrodon acutus*, *Lachesis muta*, and even from an elapid, *Dendroaspis jamesoni*.

Mambin was isolated from *Dendroaspis jamesoni kaimosae* and has a structure similar to that of the postsynaptic neurotoxins, fasciculins and cytotoxin. It has 59 amino acids and 4 bi-sulfur bridges that give it a "three-finger" configuration.

3.2.3.3 Enzyme inhibitors and activators

Several molecules exert a specific activity on certain natural enzymes of vertebrates. Generally these proteins are without any toxic or clinical effect, but none-the-less present a potential pharmacological interest.

Bothrojaracin from the venom of *Bothrops jararaca* is a double chain protein of 27 kDa, similar to botrocetin, causing aggregation of the blood platelets; it inhibits natural thrombin and the TSV-Pa activates the plasminogen.

A double chain glycoprotein of 21 kDa of *Deinagkistrodon acutus* inhibits factor X. A very similar protein but with a molecular weight of 27 kDa has been extracted from the venom of *Protobothrops flavoviridis*.

Finally the inhibitors of the enzyme that converts angiotensin I into angiotensin II need to be mentioned as they cause severe clinical problems. One of them present in the venom of *Bothrops moojeni* has been the object of many studies and has become the leader of antihypertensive drugs used therapeutically. It is possible that other inhibitors of the angiotensin conversion enzyme exist in certain venoms of the Viperidae like *Bitis arietans* in particular.

3.2.3.4 Cobra venom factor

The cobra venom factor (CVF) consists of several oligosaccharide chains and a polypeptide chain linked together by di-sulfur bridges. The functional zone of the molecule is found on the peptide chain but it appears that the presence of the saccharide chains is necessary for the CVF to function. The average molecular weight of the known cobra venom factors is 130 to 160 kDa. The CVF activates the serum complement. It acts principally on fraction 3 of the complement (C_3) and secondarily on fraction 5 (C_5).

3.2.3.5 Dendropeptin

This peptide was isolated from the venom of *Dendroaspis angusticeps* at the beginning of the 90s. It consists of 38 amino acids and a di-sulfur bridge giving the molecule the form of a ring of always 17 amino acids. By its structure and its properties dendropeptin belongs to the family of cardiac diuretic factors (atriopeptine or atrial natriuretic peptide and brain natriuretic peptide).

These peptides play a hormonal role in renal excretion and the regulation of arterial pressure. In addition dendropeptin facilitates the liberation of the second hormonal messenger (cyclic adenosine monophosphate or cyclic guanosine monophosphate) in the endothelial cells.

3.3 Variability of venoms

The variability of venoms has been reported since antiquity based on clinical considerations and recently based on experimental proof. Subsequently it has been largely confirmed by biochemical and immunological techniques.

The methodology used in the study of venom variability is of considerable importance. In effect, the techniques used can cause marked differences in the observed results. The clinical study, can show many differences in the symptomatology of envenomations caused by the same species of snake in different patients, is not exempt from this rule. In addition to the variability of the venom of the species in question, the conditions of the bite on one hand and the physiological reactions of the bitten subject on the other hand can explain very different clinical disorders. Experimental studies during which the parameters are controlled are equally prone to many artifacts. Using the venom of a single and correctly identified snake brings the necessary guarantee to the evaluation of variations in the different stages of the analysis. Mixing of venom from different animals leads on the contrary not only to a risk of an identification error or the inclusion of an "intruder", but also introduces a confusion between individual, sub-specific and specific variations in the case of species complexes, that are a frequently observed phenomenon in zoology.

3.3.1 Variations between species

The variability between different snake families is, as we have seen, extremely large. The venoms of the Elapidae are rich in toxins with a strong neuro-muscular tropism; the venoms of the Viperidae have an arsenal of complex enzymes that act on many systems, particularly coagulation. A second level of variability concerns the genus, which certainly has a large number of common chemical structures, but all the same one can observe important biochemical and immunological differences. At species or subspecies level the similarity between the venoms is still stronger. One could qualify this as phylogenetic variability.

3.3.2 Ontogenic and environmental variations

Within a population, even within a small group of related individuals, one can still find important variations in the venom which are of genetic origin as was confirmed about 20 years ago. An alkaline fraction that is absent in males has been demonstrated in females of *Crotalus adamanteus* and *Calloselasma rhodostoma*.

When one analyses the composition of the venom of an individual throughout its life, one notices a strong biochemical stability. However, some authors have described significant variations in the venom taken from one individual at different times in its life. A variation in toxicity as a function of age has been demonstrated in a single specimen of *Crotalus atrox*. The venom of young *Bothrops jararaca* has a higher toxicity for frogs than that of adults, which corresponds to the fact that the young *B. jararaca* eat frogs and lizards, but the adults don't. This phenomenon has not been observed in *B. alternatus* that eats mammals at all ages.

Certain proteins appear, disappear or become modified during a life span, particularly in the same individual which induces differences in the venom synthesized at birth and that synthesized in adulthood. In certain species the enzyme potential appears to develop during growth and to decrease in senescence. Without doubt the digestive action is more useful to adults swallowing larger prey than to juveniles. The latter, on the contrary, would derive an advantage from the immobilizing and immediate toxic action, allowing them to capture and quickly immobilize their prey. However, certain observations, like the great variations in enzyme activity in a specimen of *Pseudonaja textilis* during the year, cannot be explained.

Nutrition does not modify the composition of the venom, nor does the rhythm of the seasons, not even, as has been suggested, important physiological factors like reproduction or hibernation. The characteristics of bothrojaracin, one of the components of the venom of *Bothrops jararaca*, are constant in a given individual. The observed seasonal variations were insignificant and related to the quantity of venom and the proportion of bothrojaracin, not to its structure. The seasonal variations mentioned by some authors generally are due to methodological artifacts: at different times of the year the analyzed venoms consisted of mixtures of venoms from different individuals, apparently sometimes belonging even to different populations. With regard to clinical observations of a more pronounced severity of bites in certain seasons, they could be linked to the average age of the population, which when older would consist of larger individuals capable of delivering larger quantities of venom. This is seen particularly in countries with a moderate climate when after the end of winter the snakes emerge from hibernation. These are adults looking for food or for a sexual partner. In autumn on the contrary, the young ones are in the majority, looking for their future territories. The lower quantity of venom in these individuals, more than an improbable qualitative difference, is the probable reason for the lower severity of envenomations in autumn.

However, environmental conditions can exert a significant influence on venom composition. It has recently been shown that the biochemical variation of the venoms of distinct populations of *Calloselasma rhodostoma* was strongly correlated to the nutritional regimen of the studied populations. This could mean that the availability of certain prey species could exert a selective pressure on the composition of the venom.

The venom variations appear to relate at the same time to the concentration of the different fractions as well as to their biochemical structure. Several genetic mechanisms can explain these phenomena. The gene coding for the synthesis of a protein could consist of several alleles that lead to sufficient structural modifications to effect a significant functional variation. The metabolism of a protein is regulated under the influence of endogenous and exogenous factors, which affect the synthesis of a molecule or its expression. Finally, with the aim at efficacy, nature doubtlessly has favoured a certain redundancy of molecular structures, which would be capable of responding to a range of variable and unexpected situations. Thus it has been shown that one and the same snake can excrete several molecular variants of the same protein. The mix of several toxins in the venom of an individual would allow it to maintain a sufficient toxicity under varying conditions, like the multiplicity of prey animals whose sensitivity to each of the variants would differ. The genetic origin of the composition of the venom appears to be confirmed by the fact that the right and left venom glands of the same snake synthesize chemically strictly identical venoms, as has been demonstrated several times.

The impact of these biochemical variations can also change the immunological sites that are recognized by the antibodies used in the treatment of envenomations. Consequently the antivenom prepared from a venom could have a reduced efficacy with regard to venoms of individuals of the same species but from a different location. The random mixing of venoms from different populations of snakes is a precaution which has appeared to be insufficient, as the World Health Organization has been induced to create an international venom reference center to study this problem and to find a solution.

CHAPTER 3

The Toxicology of the Venoms

The toxicology of a venom results from the pharmacological action of its different components and from the response by the envenomed body. It is a complex individual phenomenon that necessitates a particular approach: perfectly controlled experimental conditions, repetition of trials and mathematical and statistical analysis of the observations.

1. Measuring Toxicity

The toxicity of snake venoms is evaluated by *in vivo* experiments aiming at determining the dose that causes the death of half of the tested animals (LD_{50}). The most frequently used experimental animal is the white mouse. The route of administration of the venom (or the toxin) varies: subcutaneous, intramuscular, intraperitoneal, intravenous, intracerebro-ventricular. However, the intravenous route gives the most homogeneous results; it is generally part of the standardized methods. To be reproducible and comparable the trial conditions are strictly codified. Environmental and experimental conditions can have a marked impact on the results (Table IV).

The model generally regarded as the most representative of the development of the pharmacological effects is represented by a symmetric and asymptotic sigmoid curve for the values 0 (no effect or 0% lethality) and 1 (maximum effect or 100% lethality). This method is based on the most frequent and most common dose/effect relation. The present statistical methods allow the modeling of the whole curve of toxicity by merging the three points of inflection in order to describe the whole of the phenomenon. They also allow the prediction of the confidence limits of the LD_{50} with the help of a simple calculation (Annex 1). A graphic technique further facilitates obtaining the result with the help of a log/probit transformation of the

Table IV
Factors influencing LD$_{50}$ variations

Factors	Variation
animal	
species	strong
strain	moderate
age	low
mass	low
sex	low
nutrition	unknown
season	unknown
technique	
route of injection	strong
Injected volume	non-significant
solvents	non-significant
temperature of the inoculum	moderate
ambiant temperature	moderate
venom	
species	strong
age	moderate
sex	non-significant*
geographical origin	moderate
individual variations	moderate
dessiccation of the venom	moderate
conservation of the venom	non-significant

* the variations linked to sex can be masked by individual variations

experimental data that transforms the sigmoid curve into a straight line (Fig. 15). This makes it possible to infer by direct reading the LD$_{50}$ and the confidence limits for a given risk.

Other models are based on the hypothesis that the time of survival is dose-dependent. They allow the use of all the animals in the trial in the same calculation, as if

The Toxicology of the Venoms

each group was dependent on the others and thus to reduce the total number of animals to be inoculated. Certain practical procedures allow the reduction of the confidence limits. However, they have the inconvenience to favor either the most toxic or the most rapid acting substance to the disadvantage of the other components of the venom. In effect, the principle of this method (direct observation of the time of survival) limits the time of measurement of the survival time of each successively inoculated animal to several minutes. This disadvantage probably is small in the dose-dependent toxins; it is likely to be more important with regard to the time-dependent enzymes.

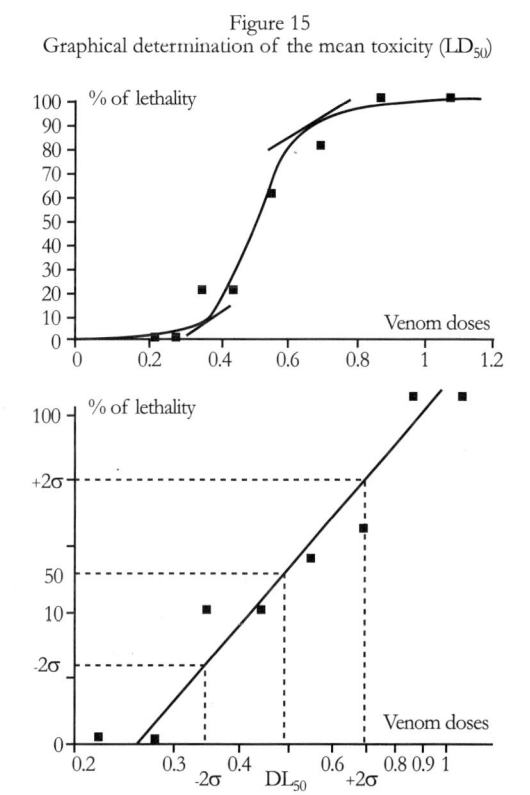

Figure 15
Graphical determination of the mean toxicity (LD_{50})

Invertebrates are used more and more often for toxicity tests (cockroaches, diptera larvae). New biological models such as isolated tissue preparations (neurons, muscle or cardiac fibers, ganglions, cell cultures) tend to replace the laboratory animals.

The World Health Organization prescribes other more specific tests for measuring in particular the defibrinating activity, the hemorrhage-causing activity and the necrotizing activity. The principle of these tests is to inject increasing doses of venom into animals and to quantify the reaction after a defined time interval. The defibrinating activity is measured by taking a blood sample and determining the residual fibrinogen or measuring the clotting time in an untreated test tube with evaluation of the quality of the clot. The haemorrhage causing and necrotizing activities are evaluated in animals as a function of the diameter of the lesions on the inside of the skin at the point of injection, 1 or 3 days after the injection of the venom. The units are defined as the quantity of venom necessary to obtain a precise reaction (10% of the initial value of fibrinogen for the minimal defibrinating dose, a lesion with a diameter of 10 mm for the minimal hemorrhage-causing or necrotizing doses).

Factors causing a variation of the toxicity depend on the venom itself, the composition of which could differ, on the envenomed animal (species, weight, size, sex, physiological state, environmental conditions) and on the experimental conditions (route of administration, speed and volume of injection, qualities of the solvent). In practice it is seen that the most important variations are linked to the venom and a lack of homogeneity of the animals used. The experimental conditions and the employed method are coincidental, with the exception of the route of injection (Table V) and perhaps certain environmental conditions like ambient temperature.

2. Effects on Cells

The venoms of the Elapidae contain a large quantity of toxins acting on the cell membranes. The cytotoxins have many properties that have led to giving them different names: cardiotoxin, direct lytic factor, membrane toxin, haemolysin, cytolysin and cytotoxin, names which they have retained.

The principal property of cytotoxins is the power to cause the lysis of the cell membranes. Different modes of action using different regions of the molecule depending on the type of affected membrane, appear to be at the origin of this definite phenomenon. For a long time the observed effects were partially due to contamination during the preparation of the extracts: neurotoxins when the isolation of cytotoxins was carried out by fractionated precipitations, phospholipases up to recently when molecular sieving and ion exchange columns were used for purification. The synergy between cytotoxin and phospholipase is remarkable and their reciprocal contamination causes a large increase in the cytolytic power of

Table V
Variations of the toxicity of venoms as a function of the route of administration

	LD_{50} (μg/mouse)				LT_{50} (minutes)	
	IV	IP	IM	SC	IV	SC
Agkistrodon contortrix	218	210		522		
Agkistrodon piscivorus	84	102		502		
Crotalus adamanteus	34	38		274		
Crotalus atrox	44	74		356		
Crotalus durissus terrificus	5,6	5		28		
Crotalus scutulatus	4,2	4,6		6,2		
Crotalus viridis	9,6	9	26			
Dendroaspis jamesoni	8,41	11,88	11,88	8,41	80	
Echis ocellatus	11,88	11,88	33,49	33,49	85	
Naja melanoleuca	7			10	40	65
Naja nigricollis	28			65	80	350
Sistrurus miliarius	58	138		502		
Vipera berus	11	16		129		
Cytotoxine	3,5			8	40	150
Toxin α	0,15			0,2	25	40

both of them. The relationship of the two molecules remains uncertain and it is still unknown whether the cytotoxins have a phospholipase activity and whether they act via an enzymatic mechanism or by fixing themselves on a particular membrane receptor.

The present hypothesis is that an electrostatic bond between cytotoxins and certain neutral or acidic phospholipids allows the cytotoxin-phospholipid complex to penetrate the carbohydrate layer of the cell membrane. The presence of the complex in the superficial layer of the cell membrane causes a weakening of the membrane via a still unknown mechanism. The cytotoxin-phospholipid binding is blocked in the presence of calcium ions and to a lesser degree magnesium ions. In any case, the increase in volume of the cell, which fills itself with water, appears to be linked to a disturbance of the ion exchanges somewhere in the membrane, which allows us to suppose that the cytotoxins act at the level of the ion channels, notably sodium. A connection between the cytotoxins and the glycoproteins of the cell membrane surface, some of which are believes to perhaps play a role of membrane receptor for cytotoxins, has not yet been proven. Only experiments carried out on membrane fragments, which are difficult to interpret, support this hypothesis.

The cytotoxins depolarize the cytoplasmic membrane of excitable cells. They activate phospholipase C which hydrolyses the triglycerides of the membrane, producing on one hand an alteration of the membrane and on the other hand an inhibition of the calcium/magnesium pump which causes a loss of calcium into the extracellular environment. The increased calcium concentration triggers a muscle contraction. Striated, smooth and cardiac muscle are all affected by this depolarizing action as well as neurons to a lesser degree. The fact that only excitable cells are lysed supports the hypothesis of a specific cytotoxin receptor on the cytoplasmic membranes of the excitable cells, but this receptor has not yet been identified.

The disintegrins inactivate the integrins, which are transmembrane heterodimerics that transmit to the cell from the outside the messages of growth, migration and differentiation and which strongly support the inflammatory response and cellular adhesion. The disintegrins are more common in the venoms of the viperids and present variable specificities. Most inhibit the $\beta 1$ and $\beta 3$ integrins and will have multiple impacts at the level of the diverse involved systems (inflammatory response, immune defense, blood coagulation, hormonal regulation) as well as consequential effects on these latter reactions (interruption of signals from the exterior, inhibition of cell adherence, blockage of platelet aggregation). The specificity of the disintegrins is linked directly to the variability of the membrane receptors that recognize them. The disintegrins of *Echis* and *Eristocophis* inhibit more particularly the cell adhesion of the immunoglobulins G and of fibronectin respectively.

3. Effects on the Nervous System

Many snakes have a venom that can quickly immobilize a prey. Probably this has partly become necessary due to the absence of legs that could help to hold the prey during its agony. The substances that carry out these functions block the nerve conductivity, preventing the transmission of the signals that enable motility. The nervous signal is propagated along the nerves by means of a short depolarization of the cell membrane. This depolarization is transmitted from one nerve to another or from nerve to muscle across a synapse, which assures the link between these different elements with the help of a neuro-mediator. The neurotoxic venoms act either before the synapse on the axon or after it directly on the post-synaptic membrane at the neuro-muscular junction or specifically on the neuro-mediator.

3.1. Transmission of the nervous signal

The muscular contraction is the consequence of the arrival of an electric wave at the level of the striated muscle fiber. The depolarization of the membrane that assures the transmission of the nervous signal along the axon, triggers the release of a neuro-mediator at the synapse which allows the signal to be transferred across the synapse so it can reach the muscle and provoke the muscular contraction.

3.1.1 Propagation of the nervous signal along the axon

At rest there is a difference in the electrical potential on the two sides of the membrane. A stimulus provokes the release of potassium out of the cell and instead the penetration into the cell of sodium. This phenomenon causes an inversion of the electrical potential called depolarization. The passage of the sodium and potassium ions across the cell membranes takes place through specific channels (Fig. 16).

These consist of protein groups the structure of which is modified on arrival of a variation of the electrical potential of the membrane of a predetermined value, therefore their name voltage dependant ion channels. The transfer of the ions is an active mechanism and necessitates a minimal amount of energy. The sodium channel is activated more rapidly than the potassium channel, which explains the

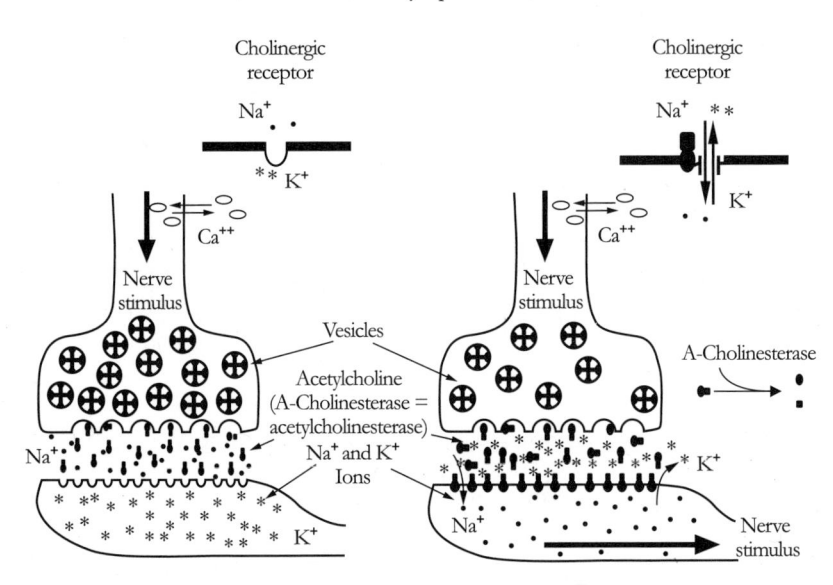

Figure 16
Neuro-muscular synapses and the nerve stimulus

amplitude and the brevity of the depolarization. As it travels along the ion exchange it allows the nervous signal to be propagated along the axon in the form of a wave called action potential.

On reaching the end of the neuron, the depolarization of the presynaptic membrane allows the penetration of calcium into the cell due to the intervention of a particular voltage-dependant channel, and this triggers the release of acetyl choline into the inter-synaptic space or synaptic gap.

3.1.2 Crossing the inter-synaptic space

The synapse presents itself as an interruption between two neurons or between a neuron and the muscle that it governs (cf. Fig. 16). The acetyl choline released into the synaptic gap establishes a brief contact with a specific receptor on the post-synaptic membrane. The cholinergic receptor consists of five subunits each with variable affinities for different substances. This receptor is itself connected to an ion channel allowing the transfer of sodium and potassium from one or the other side of the postsynaptic membrane. In this way the depolarization of the postsynaptic membrane allows the propagation of the action potential and the contraction of the muscle. These channels are called chemo-dependent channels as they are sensitive to the action of a particular molecule. Two types of receptors are identified by their response to particular substances. The nicotine receptors which have an altogether relative homogeneity, are found in the ganglions of the vegetative nervous system, the motor plates of the skeletal muscles, the medulla of the adrenal gland and some zones of the central nervous system (spinal cord and the optic roof essentially); the muscarin receptors respond to muscarin, an alkaloid of certain fungi (the toadstool for example) and are widely distributed in the brain and in the ganglions of the parasympathetic nervous system. Their activation leads to an ion exchange: sodium ions penetrate and potassium ions leave the cell. According to their function one distinguishes presently nicotine and muscarine receptors.

After the activation of the cholinergic receptor acetylcholine is hydrolysed by an enzyme that is present in the synaptic gap: acetylcholinesterase. Then the receptor finds itself free, ready to receive a new molecule of acetylcholine. Certain synapses respond to stimulation by other chemical mediators: adrenaline, γ-amino butyric acid or GABA in mammals and glutamate in arthropods.

3.2. Snake venom effects on nerve conductivity

The neurotoxins of snakes are classified in four groups according to their point of intervention and their mode of action.

3.2.1 Postsynaptic toxins (α-toxins)

They bind specifically to a chemo-dependent receptor and disturb its functioning: the first that were discovered were the curarizing toxins which bind to the cholinergic receptor of the postsynaptic membrane. They act in the same way as curare, an alcaloid extracted from strychnos, South American plants.

The diversity of cholinergic receptors in the nervous system with regard to their structure and to their localization in the nervous system allows two types of neurotoxins to be distinguished: the nicotine-like toxins which selectively recognize the nicotine receptors of acetyl choline and the muscarine-like toxins which bind to muscarine receptors and were discovered more recently.

Toxins binding to adrenergic receptors are also found in snake venoms, but they are rare.

However, a toxin with an action on GABA or glutamate-dependent receptors has not yet been isolated from snake venoms. Such toxins occur nonetheless widely in the animal world, notably in scorpions, spiders and mollusks.

3.2.1.1 Nicotine-like toxins

The molecular structure of nicotine receptors varies according to their distribution in the nervous system. The α-neurotoxins have a stronger affinity to the nicotine receptors of acetylcholine in the peripheral neuro-motor plates than to other ones.

Short and long neurotoxins have an identical mode of action. Their site of binding on the cholinergic receptor of the neuro-motor plates of the striated muscle fibers (motor muscles) is the same and it induces the closure of the ion channel (Fig. 17).

Figure 17
Mode of action of snake neurotoxins

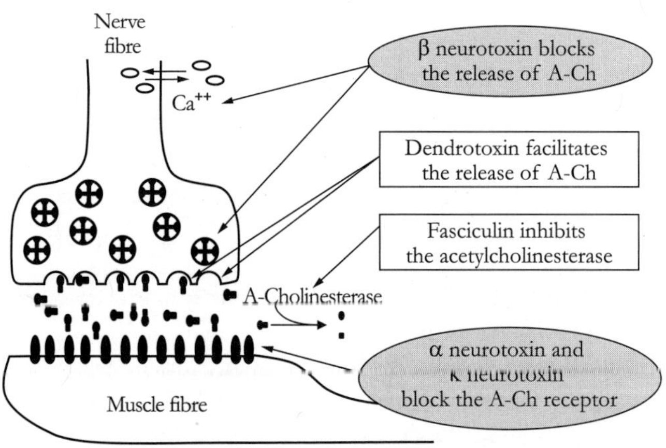

The nervous flux is blocked which causes a flaccid paralysis identical to the one seen in curarization. The strong binding of the neurotoxin to its receptor explains the persistence of the paralysis and the difficulty habitually encountered to remove the neurotoxin from the receptor. Between the short and the long neurotoxins there is a clear difference in their affinity to the cholinergic receptor. The long neurotoxins bind and dissociate themselves more slowly than the short ones.

The toxic site of the curarizing neurotoxins, the site of their binding with the receptor, is concentrated at the level of a dozen amino acids close to each other when the molecule takes on its tri-dimensional shape (cf. fig. 14). The aspect of this region of the molecule is incidentally very similar to that of the curare molecule. The epitopes, the regions of the molecule to which the antibodies attach are close to the toxic site in some and in distant regions for others. Thus one can envisage two types of antibodies, "protective" and "curative" antibodies and their clinical and therapeutic importance will be shown later.

The κ-neurotoxins have in contrast to the α-neurotoxins a high degree of affinity to the cells of the parasympathetic (ciliary ganglions) and sympathetic ganglions (lumbar ganglions). The binding to the central nicotine receptors (cervical ganglion, retinal ganglion, cerebellum, corpus striatum) is equally strong though slightly les pronounced. This selectivity confirms that there are several types of nicotine receptors.

The action of the κ-neurotoxins is translated into a blockage of the cerebellar (motor co-ordination, equilibrium and orientation), cortical, particularly of the prefrontal cortex (capability to move) and optic functions.

3.2.1.2 Muscarin-like toxins

They have the property of binding to the muscarin receptor of acetyl choline. The specificity of the muscarin-like toxins for the receptors of the type m_1, essentially present in the central nervous system, is stronger and sometimes even exclusive, than for the receptors of the type m_2 that are found in certain organs like the heart, or of the sympathetic nervous system.

3.2.1.3 Adrenergic toxins

Vipoxin is one of the rare known adrenergic toxins extracted from snake venoms. It has no action on the nicotine and muscarine receptors. It shows an equal affinity for the a_1 and a_2 adrenergic receptors but does not act on the b receptors. Injected intravenously it does not present any toxicity. However, when injected intracerebrally into an animal, it causes convulsions.

3.2.2. Presynaptic paralyzing toxins (β-toxins)

Their paralyzing action is exerted by inhibiting the release of acetyl choline. They are present in the venom of many elapids and of certain viperids. Their mode of action is less well known than that of the curare-like neurotoxins.

Depending on their specific origin they consist of one, two or four subunits and all have in common a phospholipase-like action that is necessary for their toxic activity. The inhibition of the phospholipase-like action of the β-toxines by one or other process leads to the loss of toxicity of the toxin. However, the mode of action of the β-toxins on the membrane and on the receptor to which they bind appears to be variable, depending on which one is considered. This is because of the own specificity of each toxin as well as of the presence of different receptors in the tissues of the animals used in the experiments. β-Bungarotoxin acts on the voltage dependant potassium channel while crotoxin binds to a not yet identified protein membrane.

The β-toxins cause similar neuro-physiological problems at the end of the motor nerves, although they do not intervene on the same receptors as the α-toxins. At lethal dose the β-toxins cause a muscular paralysis by stopping the transmission of nervous signals. As with the curare-like toxins or α-toxins death is induced by the paralysis of the respiratory muscles. However, the neurophysiological mechanism appears to be very different. The progressive blockage of acetylcholine release under the action of β-toxins occurs in three stages. After a rapid diminution of acetylcholine release by the pre-synaptic membrane, one can observe a brief phase of recovery before a complete and definitive drop. The first phase is independent of the phospholipase activity of the β-toxin and would be due to the direct action of the toxin on the membrane receptor. In mammals the second phase is equally independent of the phospholipase action of the β-toxin and would indicate a partial repolarization of the membrane that would allow the liberation of small quantities of acetyl choline in the synaptic gap. In frogs this second phase depends on the phospholipase function, which seems to indicate a different mode of action linked to a hydrolysis of the cytoplasmic membrane, even though its neurophysiological translation is the same. During the third phase, finally, the catalytic activity on the phospholipids of the membrane leads to the definitive arrest of ion exchanges from one side of the neuron to the other.

The first two phases suggest on one hand binding to a membrane receptor and on the other hand an intervention on an ion channel, which may very well differ with the toxin (probably the potassium channel for β-bungarotoxin). In addition the catalytic action of the phospholipase A_2 on the architecture of the muscle fibre is seen as a muscle necrosis that varies in importance with the toxin. It is very high in the toxins of marine and Australian elapids and is characterized by an increased myoglobinuria that sometimes causes renal insufficiency.

Two toxins present in the venoms of *Dendroaspis* can be included in this group: calciseptin and calcicludin that act on the receptor of the calcium channel and inhibit its physiological functioning. However, their toxicity and their mode of action are still poorly known and it is likely that their role in the envenomation is negligible, at least in humans.

3.2.3 Pre-synaptic facilitating toxins

The facilitating toxins act at the same time at the pre-synaptic level and facilitate the release of acetyl choline (cf. Fig. 17).

The 50% lethal dose (LD_{50}) that can kill half of the trial animals, is about 2000 times higher than that of β-bungarotoxin, a paralyzing pre-synaptic neurotoxin, which has a non-phospholipase sub-unit that is structurally homologous to the dendrotoxins. In any case, injected intracerebrally the dendrotoxins are as toxic as most of the known β-toxins.

The dendrotoxins show an elevated affinity to certain voltage-dependent potassium channels. They facilitate the release of acetyl choline in the synaptic gap and cause a strong depolarization of the postsynaptic membrane. The precise mode of action at the level of the pre-synaptic membrane has not been elucidated. It has been shown that the dendrotoxins remain only a few minutes in contact with their target and that their effect continues long after their elimination.

The symptoms include convulsions, epileptic disorders, demonstrating a direct action on the central nervous system, and respiratory paralysis, the true cause of death, which originates peripherally. This complex symptomatology clearly proves that that the dendrotoxins act on widely distributed receptors in the nervous system. The weak toxicity of these toxins during natural envenomations limits their clinical interest.

3.2.4. Fasciculins

They inhibit acetylcholinesterase and prevent the destruction of acetylcholine in the synaptic gap. Through this they allow an increase of the acetylcholine concentration in the synaptic gap. The cholinergic receptor is thus stimulated permanently by the acetylcholine that detaches itself causing a repeated depolarization of the postsynaptic membrane. This is translated into strong, short and repeated muscle contractions, or muscular fibrillation (fasciculation, from which the name for these neurotoxins derives). In parallel the cholinesterase inhibition effect on the parasympathetic system is seen as well. This symptomatology can persist for hours. At a higher dose the respiratory muscles become paralyzed which can be retarded significantly by high doses of atropine (50 mg·kg^{-1}). This effect on the neuro-muscular synapses could be connected to an action on the voltage-dependent potassium channels.

The presence in the same venom of dendrotoxins and fasciculins, the actions of which complement each other, illustrates the synergisitic effect that certain toxins can have amongst each other. While the former induce a rise in the release of acetylcholine in the synaptic gap, the latter enhance the duration, which in consequence would lead to the muscle contractions and their persistence. In spite of a moderate individual toxicity the morbidity-causing effect is considerable.

3.2.5 Myotoxins

The myotoxins of viperids and colubrids inhibit the potassium or calcium channels without altering the acetylcholine receptors. It is possible, at least in certain cases, that the myotoxins act by their necrotizing ability, as the potassium channels in particular cannot function if the muscle fiber is not intact; this would therefore be a non-specific effect.

4. Effects on the Cardio-vascular System

The action of venoms on certain functions can be complicated secondarily by a cardio-vascular attack. The inflammatory syndrome or haemorrhage problems provoke a lowering of the blood volume, which has repercussions on the blood pressure and on the heart. The central and peripheral neuro-muscular attacks and the respiratory impairment would equally have important cardio-vascular consequences, if only by variations of the partial pressure of gases in the blood.

4.1. Hydrolysis of the endothelium

The haemorrhagins are zinc-dependent metallo-proteases that act directly on the vascular endothelium. These enzymes share a common physiological action but are not structurally homogenous. They are found in most venoms of viperids as well as in the venom of certain Australian elapids and of colubrids. They attack the basement membrane of the endothelium. This causes on one hand an immediate extravasation of blood, which escapes from the blood vessels causing an edema or a rash, and on the other hand an activation of the physiological blood coagulation independent of the direct action of the venoms on blood coagulation.

The action of the haemorrhagins can be focal and cause bleeding at the point of the inoculation of the venom, or systemic and produce distant haemorrhages, contributing to a haemorrhagic syndrome.

The non-specific cytolysis, whether due to enzymes or to cytotoxins, can equally affect the fibers of the myocardium and the vascular endothelia, and thereby cause destruction and disturbances of varying intensity.

4.2. Effects on blood pressure

The action of the venom can induce a state of shock via two mechanisms.

- Certain peptides inhibit the conversion enzyme which hydrolyses angiotensin I into angiotensin II with a double consequence: on one hand angiotensin II disappears and no longer excercises its vasoconstrictive activity, as well as that of aldosterone with its role of sodium retention, and on the other hand the levels of the powerful vasodilators bradykinine and prostaglandins increase. The consequence of these mechanisms is a severe drop of peripheral resistance causing the arterial pressure to drop as well. This drop is early (a few minutes after the inoculation of the venom), fast and severe. The heart rhythm is not modified and this excludes a direct cardiac toxicity. This phenomenon is observed with the venoms of certain viperids like the *Bothrops* spp. in South America and *Bitis arietans* in Africa.

- Also a cardiogenic shock can be encountered with venoms that contain a cardiotropic toxin like the sarafotoxins of the venoms of *Atractaspis* or taicatoxin found in the venom of *Oxyuranus scutellatus*.

The mode of action of the sarafotoxins is largely described by reference to the endothelins, hormones that play an essential role in the local regulation of blood pressure. The sarafotoxins cause the reversible contraction of the smooth muscles with a marked preference for certain organs depending on the sarafotoxins under consideration. The sarafotoxins bind to cellular receptors; the activation of phospholipase provokes the hydrolysis of phosphoinositide, a metabolite of arachidonic acid, and the release of prostaglandins. Simultaneously they cause the cellular uptake of calcium. The sarafotoxins increase the force of contractions of the heart muscle fibers (positive inotropic action) without disturbing the rhythm in the beginning, but later can evolve towards a slowing down of the auriculo-ventricular conduction (1st degree auriculo-ventricular block, namely an increase in the PQ interval). The disturbance of the conduction affects the electrical command centers of the heart (notably the auriculo-ventricular node) and leads to an arythmia most frequently in the form of a ventricular bradycardia and a desynchronization of cardiac excitation.

Taicatoxin blocks the calcium channel of the heart muscle fiber and this leads to a sinusoid bradycardia with an atrio-ventricular block.

Dendropeptin is a direct competitor of atriopeptin. This hormone encourages the renal excretion of sodium and has a vasodilatory activity; it inhibits aldosterone. It appears that its impact on arterial pressure is modest.

5. Effects on Haemostasis

On one hand the impermeability of the vascular system is assured by the vascular walls consisting of fibers and more or less disjointed endothelial cells and on the other hand by blood coagulation which limits the fluidity of the blood and its running out of the blood vessels.

Snake venoms, particularly those of the viperids, interfere with all the mechanisms of coagulation. The role of haemostasis is to prevent haemorrhages and schematically it includes three intimately interconnected periods (Fig. 18).

Figure 18
Stages of haemostasis and of fibrinolysis

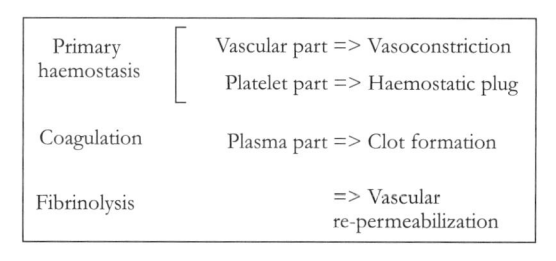

Primary haemostasis	Vascular part => Vasoconstriction
	Platelet part => Haemostatic plug
Coagulation	Plasma part => Clot formation
Fibrinolysis	=> Vascular re-permeabilization

The first or vascular step leads to a reflex-induced vasoconstriction and makes the vascular wall accessible to the plasma factors and this stimulates the following steps of coagulation. This is followed by the platelet step which allows the aggregation of the platelets and the formation of the platelet plug around which the thrombus will be formed. During the third step or the actual coagulation, the thrombus forms and solidifies with the help of a fibrin net that holds the blood cells in place (fibrin formation) and consolidates the platelet plug. The fibrin net organizes itself under the action of an enzyme, thrombin, which is present in the blood in its inactive form: prothrombin. Its activation (thrombin formation) sets in motion the cascade of enzymatic reactions that allow the blood to coagulate.

After that the physiological destruction of the thrombus or fibrinolysis allows the blood vessel to become permeable again. While the three steps of haemostasis take place within a few minutes, at most 20 or 30, fibrinolysis occurs much later around the third day and this allows the vascular endothelium to repair itself.

5.1 The physiology of haemostasis

In the normal state blood coagulation is the result of an ensemble of biochemical reactions that can take place simultaneously or successively. Classically, blood coagulation is primed by a vascular wound. However, in practice the activation of an enzyme complex belonging to any of the steps of haemostasis can be induced by various general or localized circumstances.

5.1.1 Platelet aggregation

The adhesion of the platelets or thrombocytes amongst each other and to the vascular endothelium is provoked by a lesion of the latter. Von Willebrandt's factor is a protein that recognizes both, the collagen and the receptors of the thrombocyte membrane and it assures a covalent bond between these different substrates and this leads to the aggregation. The main membrane receptor is a glycoprotein, GPIb, which does not need a specific activation. A second receptor, GPIIbIIA, is an integrin, a transmembrane protein, which needs to be activated first by adenosine di-phosphate (ADP) and thrombin.

Once agglutinated the thrombocytes release their granules which contain vasoconstricting factors (serotonin, platelet growth factor or PDGF, thromboxane A_2, which also mediates inflammation) and factors that support cellular adhesion and thus platelet aggregation (fibronectin, von Willebrandt's factor, fibrinogen, ADP which also mediates inflammation).

5.1.2 Blood coagulation

5.1.2.1 Thrombin formation

The formation of thrombin depends on the prothrombinase, which transforms prothrombin into thrombin. Prothrombinase is an enzyme complex consisting of the activated factor X (= Stuart factor) together with factor V (= pro-accelerin). For exerting its catalytic activity the complex needs to bind onto phospholipids of tissue or platelet origin in the presence of calcium ions. Prothrombinase breaks the specific peptide bonds of the prothrombin molecule (Fig. 19) and successively allows the release of prethrombin 2 and then after a second hydrolysis of natural thrombin (α-thrombin). The latter is a proteolytic enzyme of the serinproteases group that causes the conversion of fibrinogen to fibrin during fibrin formation.

Figure 19
Hydrolysis sites of the prothrombin molecule (after Samama, 1990)

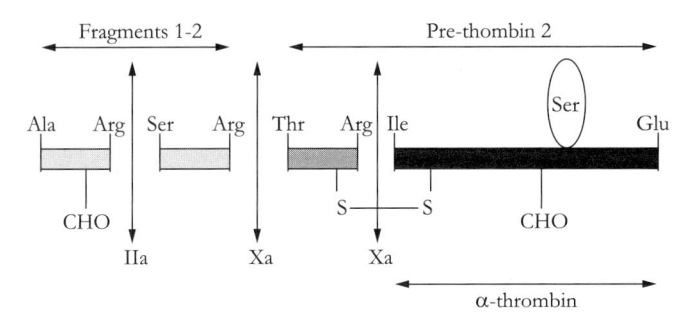

```
     Fragments 1-2              Pre-thombin 2
  ←──────────────→   ←──────────────────────────────→
                                              ┌Ser┐
 Ala  Arg  Ser  Arg  Thr  Arg  Ile            └───┘      Glu
      ▓▓▓       ▓▓▓       ▓▓▓       ████████████████
       │         │         │         │
      CHO        │        S─┼─S      CHO
       ↓         ↓         ↓
      IIa        Xa        Xa
                        ←──────────────────────────→
                                  α-thrombin
```

5.1.2.2 Fibrin formation

Fibrinogen is a large glycoprotein of 340 kDa and consists of six protein chains assembled in symmetric pairs. Thrombin releases two pairs of small peptides, the fibrinopeptides A and B (Fig. 20) triggering the polymerization between them and the formation of a net of soluble fibrin (Fig. 21).

The release of the fibrinopeptide A happens quickly; in the place left vacant the monomers polymerize longitudinally in the form of "protofibrils". The fibrinopeptide B detaches itself from the protofibrils more slowly and this allows a lateral polymerization that leads to the formation of thicker fibers: the soluble fibrin. The

Figure 20
Transformation of fibrinogen to fibrin monomers (after Samama, 1990)

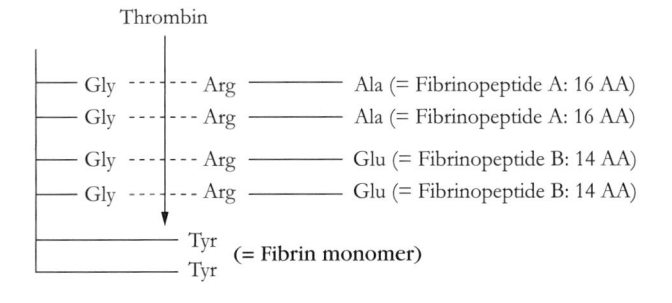

```
        Thrombin
           │
    ┌─ Gly ─┼── Arg ──────── Ala (= Fibrinopeptide A: 16 AA)
    ├─ Gly ─┼── Arg ──────── Ala (= Fibrinopeptide A: 16 AA)
    ├─ Gly ─┼── Arg ──────── Glu (= Fibrinopeptide B: 14 AA)
    └─ Gly ─┼── Arg ──────── Glu (= Fibrinopeptide B: 14 AA)
            ▼
          Tyr
    ┌─────────    (= Fibrin monomer)
          Tyr
```

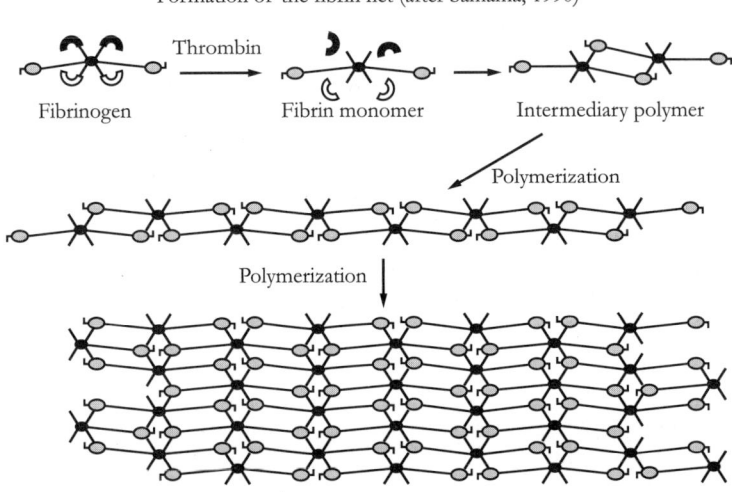

Figure 21
Formation of the fibrin net (after Samama, 1990)

activity of the factor XIII or fibrin stabilizing factor (= FSF) leads to an insoluble fibrin which forms a stable clot. The insoluble fibrin is obtained by transamination, the formation of new chemical bonds between the monomers of the polymerized fibrin.

5.1.2.3 Fibrinolysis

The stabilized clot persists for about 72 hours before being degraded by plasmin. Under normal conditions the fibrinolysis has to be regarded as a regulatory system that assures an equilibrium with coagulation.

Plasminogen binds to the fibrin during the formation of the latter. At its release the tissue activator of plasminogen binds equally to the fibrin and hydrolyses the plasminogen that transforms itself into plasmin. Plasmin is thus included in the thrombus and thereby protected against physiological inhibitors. The latter can intervene to neutralize it only after the destruction of the clot. The activity of plasmin is exerted on fibrinogen, soluble fibrin and stabilized fibrin. Only the hydrolysis of the latter generates the specific fibrin degradation products (FDP_n) that are different from the products of fibrin degradation obtained after the digestion of fibrinogen or of the soluble fibrin (FDP$_g$). The difference between FDP$_n$ and FDP$_g$ is difficult to establish with the help of current techniques. It is due to hydrogen bonds and covalent bonds that characterize the stabilized fibrin and are resistant to the activity of plasmin. This distinction becomes important in cases of envenomation and their treatment.

5.1.3 Mechanisms regulating haemostasis

In addition to the physiological fibrinolysis which has been described above and which has a slow action after the stabilization of the thrombus, the control of the different chemical reactions falls under complex auto-regulating mechanisms.

5.1.3.1 Local regulation

Membrane phospholipases form a privileged interface limiting the phenomenon to the injured area and reducing the risk of its extension.

5.1.3.2 Systemic regulation

Most of the coagulation factors exert inhibiting effects on the reactions further upstream and this allows the slowing down of the coagulation process. In vertebrates two principal mechanisms assure the regulation of coagulation in the long term: the serin-proteases system of inhibitors (serpins = serine protease inhibitors with anti-thrombin III in the leading role) and protein C, which is induced by thrombin in the blood circulation. Protein C hydrolyses on one hand the activated factors VIII and V thus blocking the physiological reactions of the coagulation and on the other hand the inhibitor of the plasminogen activator that triggers fibrinolysis as described above.

5.2 The mode of action of snake venoms on haemostasis

Each venom contains an ensemble of substances that either enhance or inhibit the coagulation at several levels (Fig. 22). Whichever the mode of action, this is often translated by a clinical haemorrhagic syndrome, sometimes sudden, more often progressive and possibly with a fatal outcome.

5.2.1 Coagulation factors

Although several mechanisms of action have been identified, the most general case can be schematized by invoking the principle of substitution: a protein present in

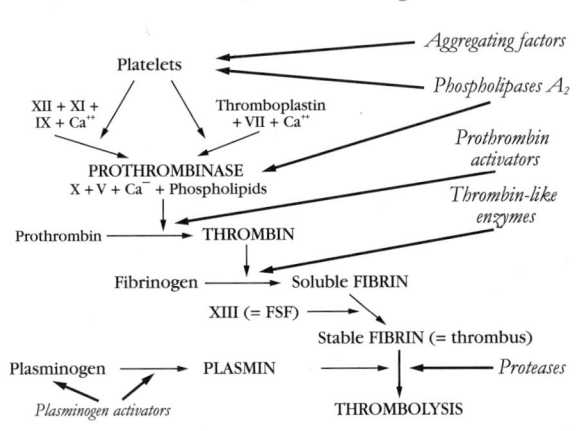

Figure 22
Snake venom actions on blood coagulation

```
                                              ─── Aggregating factors
                  Platelets ◄──────────
                     ▲    ▲  ◄─────────────── Phospholipases A₂
  XII + XI +       ╱        ╲
  IX + Ca⁺⁺      ╱     Thromboplastin                 Prothrombin
                ╱       + VII + Ca⁺⁺                  activators
               ▼            ▼
            PROTHROMBINASE  ◄──────────
            X + V + Ca⁻ + Phospholipids          Thrombin-like
                                                   enzymes
    Prothrombin ──────► THROMBIN  ◄──────────
                           │
                           ▼
            Fibrinogen ──────► Soluble FIBRIN
                       XIII (= FSF) ─────►
                                          ╲
                                    Stable FIBRIN (= thrombus)
   Plasminogen ─────► PLASMIN ──────►    │   ◄───── Proteases
        ▲        ▲
        │        │                    THROMBOLYSIS
   Plasminogen activators
```

the venom has analogue properties to one of the coagulation factors and takes its place. When the coagulation process is activated (process of disseminated intravascular coagulation or DIC syndrome) it persists until one or several coagulation factors have been used up, consumed (= consumption phenomenon) and leads to a haemorrhagic syndrome due most often to an afibrinogenaemia. One can oppose two large groups of coagulation factors: the activators of prothrombin and the thrombin-like enzymes. The activators of the other coagulation factors, notably the activators of factor V, appear to be secondary from a clinical point of view.

5.2.1.1 Activators of factor V

Several venoms of viperids, notably *Daboia russelii*, *Vipera aspis* and *Bothrops atrox*, contain proteins that activate factor V and sometimes other coagulation factors. It is generally the serin-proteases that cause a controlled hydrolysis of factor V. The factor V activator extracted from *D. russelii*, RVV-V, acts on a site that is equally the target of thrombin. However, in contrast to thrombin the RVV-V is not inactivated by antithrombin III, even in the presence of heparin. Although the structure of thrombocytin, extracted from the venom of *B. atrox*, is very close to that of factor V activator extracted from the venom of *D. russelii*, it has a much wider range of activities than the latter: it hydrolyses not only factor V, but also factor XIII, factor VIII, prothrombin and fibrinogen and activates the thrombocytes.

The importance of these activators appears to vary from one venom to another. They have been discovered through their action mechanisms and their additional activities and these constitute their principal interest. The isolated activation of factor V, particularly in the absence of phospholipase and of calcium, generally leads to a modest clinical manifestation.

5.2.1.2 Prothrombin activators

Normally prothrombin is hydrolysed at the site of two peptide bonds (Arg_{271}-Thr_{272}) and then (Arg_{320}-Ile_{321}) by the activated factor V, namely bound to factor V and to a phospholipid and in the presence of calcium. This reaction releases natural thrombin or α-thrombin (cf. Fig. 19). The prothrombin activators are ranged in five groups.

- The first group consists of the majority of prothrombin activators encountered in snake venoms and a typical representative is ecarin extracted from the venom of *Echis carinatus*. They hydrolyze the bond (Arg_{320}-Ile_{321}) only and this leads to the formation of meizothrombin that has different properties from those of α-thrombin. Also this activator is independent of any co-factor and does not need factor V, phospholipids or calcium. The prothrombin activators isolated from the venoms of *Trimeresurus gramineus* and of *Bothrops jararaca* belong to this group, as well as those from most of the venoms of colubrids: *Dispholidus typus, Thelotornis kirtlandii, Rhabdophis tigrinus*.

- The activators of the second group hydrolyze prothrombin at the same sites as the physiological prothrombinase complex: the result is thus a natural thrombin. The efficacy of these prothrombin activators is multiplied by a factor of 1 to 19 million in the presence of phospholipids, of the activated factor V and of calcium. The representative of this group is the prothrombin activator of *Notechis scutatus*.

- The third group consists of the activators from certain venoms of Australian elapids like the one of *Oxyuranus scutellatus* that has the same hydrolysis site and does not require the presence of factor V but only of phospholipids and calcium.

- Group 4 consists of prothrombin activators that transform prothrombin to prethrombins 1 and 2, substances which do not have any enzymatic activity and therefore are not toxic. Acutin extracted from the venom of *Deinagkistrodon acutus* represents the type. It is found in the venoms of American and Asian crotalids (*Crotalus, Agkistrodon* and *Calloselasma* respectively).

- Finally carinactivase, an enzyme isolated from the venom of *Echis carinatus*, presents a composite structure combining a metallo-protease of the type of ecarin (group 1) with a di-meric peptide that allows the presentation of the enzyme to its substrate and facilitates its hydrolysis.

Clinically it is not always possible to differentiate between the effect of the prothrombin activator and that of the other factors acting on the coagulation, at least with the venoms of viperids, notably *Echis carinatus*. However, in certain African colubrids (*Thelotornis kirtlandii*) or in Australian elapids (notably *Notechis scutatus, Pseudonaja textilis* and *Oxyuranus scutellatus*) the prothrombin activator appears well to be principally responsible for the sometimes severe haemorrhagic syndromes seen by clinicians.

With regard to therapy it has to be noted that the prothrombin activators generally are not sensitive to the normal serin-protease inhibitors. Ecarin is inhibited by EDTA and 2-mercaptopurinol, but not by trypsin inhibitors nor by di-isopropylfluorophosphate, iodoacetate or parachloromercurobenzoate.

5.2.1.3 Thrombin-like enzymes
Very many snake species have enzymes that can transform fibrinogen to fibrin (Table VI; Fig. 23).

However, none is identical with natural thrombin, not even with meizothrombin. This is seen already in their structure. While prothrombin consists of two peptide chains with a di-sulfur bridge all thrombin-like enzymes extracted from snake venoms consist of a single polypeptide chain. In addition depending on the species the site of hydrolysis (allowing the release of fibrinopeptides A or B), the action

Table VI
Principal thrombin-like enzymes of snake venoms

Snake	Thrombin-like enzyme	heparine	hirudine	others
Agkistrodon bilineatus	bilineobin	+	0	DFP – PMSF – AT III
Agkistrodon contortrix	venzyme	+	0	PMSF
Bitis gabonica	gabonase	0	0	PMSF
Bothrops alternatus	balterobin		0	PMSF
Bothrops asper		0		DFP
Bothrops atrox	batroxobin (Reptilase®)	0	0	benzamidine
Bothrops atrox	thrombocytin	0	0	DFP
Bothrops insularis		+		PMSF
Bothrops jararaca	bothrojaracin			
Bothrops jararaca	bothrombin	+/-		DFP
Bothrops jararacussu	D-V			
Bothrops lanceolatus	F-II-1a	0		PMSF
Bothrops moojeni	batroxobin (Reptilase®, Défibrase®)	0	0	PMSF
Calloselasma rhodostoma	ancrod (Arvin®, Viprinex®)	0	0	DFP – PMSF – AT III
Cerastes cerastes	cerastocytin			DFP – PMSF
Cerastes vipera	cerastobin			DFP
Crotalus adamanteus	crotalase	0		DFP
Crotalus durissus terrificus				DFP – PMSF
Crotalus horridus		0		PMSF
Deinagkistrodon acutus	acuthrombin	0		DFP – EDTA – PMSF
Deinagkistrodon acutus	acutin	0		DFP
Gloydius blomhoffi	brevinase			
Gloydius halys	pallabin			DFP – PMSF – AT III
Gloydius ussuriensis	calobin			
Lachesis muta		0	0	DFP

Table VI, continued
Principal thrombin-like enzymes of snake venoms

Snake	Thrombin-like enzyme	Confirmed inhibitors		
		heparine	hirudine	others
Lachesis stenophrys	gyroxin	0	0	DFP – EDTA – PMSF
Ovophis okinavensis	okinaxobin	0	0	DFP – PMSF
Protobothrops flavoviridis	flavoxobin		+	DFP – PMSF
Trimeresurus elegans	elegaxobin			
Trimeresurus gramineus	grambin	0		DFP – leupeptin – PMSF
Trimeresurus stejnegeri	stejnobin			DFP – PMSF

DFP = diisopropylfluorophosphate
PMSF = Phenylmethylsulfonylfluoride
AT III = antithrombin III

Figure 23
Snake venom actions of the formation of the fibrin net

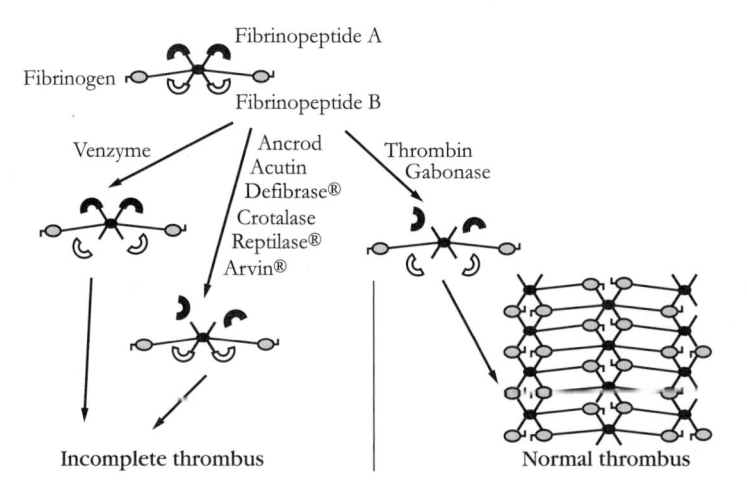

on the other coagulation factors (factors V, VIII, X, prothrombin, fibrin stabilizing factor, platelets) and the sensitivity to various natural or therapeutic inhibitors (hirudin, heparin alone or together with antithrombin III, di-isopropylfluorophosphate, phenylmethylsulphonylfluoride) are very different. The thrombin-like enzymes are placed into three groups according to the function of the released fibrinopeptide.

- Most of the known thrombin-like enzymes release only fibrinopeptide A. The fibrin which is constituted rapidly by longitudinal polymerization (fibrin I) cannot be stabilized by factor XIII or fibrin stabilizing factor (FSF), which cannot exert its action on the incomplete fibrin: therefore the thrombus remains friable.

- The thrombin-like enzymes of the second group hydrolyze fibrinogen at the level of the fibrinopeptide B. The resulting fibrin II has a lateral polymerization only and is not more stable than fibrin I and the thrombus also remains friable. The enzyme produced by the North American crotalid *Agkistrodon contortrix* as well as the enzyme extracted from the Asian crotalid *Gloydius halys* belong to this group.

- The enzymes of the third group release simultaneously the fibrinopeptides A and B. Gabonase from the venom of the African viperid *Bitis gabonica* leads to a normal polymerization of fibrin and a stable thrombus. However, in common with the other thrombin-like enzymes gabonase cannot be linked to the thrombin since it has none of the coagulation factors on which the thrombin acts (factors II, V, VIII) and it has no effect either on protein C or on the platelets.

Some rare thrombin-like enzymes have an agglutinating action on the thrombocytes. Cerastobin, a serin-protease isolated from the venom of the small North African and Near Eastern viperid *Cerastes vipera*, has simultaneously an enzymatic activity leading to the hydrolysis of fibrinogen and a degradation of the proteins of the thrombocyte membrane.

In contrast to natural thrombin most of the thrombin-like enzymes are not inhibited by heparin or hirudin, which considerably reduces the interest of these substances in the treatment of viperine envenomations. However, almost all the thrombin-like enzymes from snakes are inhibited by di-isopropylfluorophosphate. Also generally there is an antithrombin II protecting effect associated with the antivenom-serum in experimental envenomations by certain viperids like *Calloselasma rhodostoma* which has a venom particularly rich in thrombin-like enzymes.

5.2.2 Direct anticoagulant activity

The anticoagulant activity of snake venoms uses four different modes of action.

5.2.2.1 Activation of protein C

The venoms of several viperid species, notably of the genera *Agkistrodon, Bothrops* and *Cerastes*, contain protein C activators. The activation of protein C is 15 times more rapid with the activator extracted from the venom of *Agkistrodon contortrix*, venzyme (Protac®), than with its natural activator, thrombin. It takes place independently of thrombomodulin, a protein of the cytoplasmic membrane and indispensable for the physiological activation of protein C, and in the absence of calcium. However, it requires the presence of vitamin K. The protein C activator isolated from the venom of *A. contortrix* is a serin-protease of around 40 kDa and is inhibited by phenylmethylsulphonylfluoride and antithrombin III. The activation of protein C by snake venoms inhibits coagulation by hydrolyzing the activated factors VIII and V and triggering fibrinolysis by degrading the inhibitors of the plasminogen activator.

5.2.2.2 Fibrinolysis

Snake venoms contain on one hand proteolytic enzymes which have properties similar to those of plasmin and are capable of hydrolyzing fibrinogen and fibrin and on the other hand serin-proteases which activate plasminogen and further the release of natural plasmin. One of them, isolated from the venom of *Trimeresurus stejnegeri*, reproduces exactly the cleavage of plasminogen by plasminogen activators as seen in man. It is also known that the hydrolysis of soluble fibrin is more complete and rapid than that of stabilized fibrin. The thrombus produced by the action of the thrombin-like enzymes of viperid and crotalid venoms is no exception. This explains why the thrombi seen clinically in cases of viperid envenomations are friable.

Most of the proteolytic enzymes isolated from the venoms of viperids or crotalids have a predominantly fibrinogenolytic activity and behave like thrombin-like enzymes able to hydrolyze fibrin (Table VII). Some are clearly fibrinolytic, notably those present in the venoms of *Agkistrodon* (American crotalids), of *Deinagkistodon*, *Trimeresurus* or *Protobothrops* (Asiatic crotalids). Many of these enzymes are inhibited by EDTA and some are sensitive to the normal protease inhibitors (ϵ-amino-caproic acid, di-isopropylfluorophosphate, phenylmethylsulphonylfluoride, di-thiothreitol).

The role of the plasminogen activators released by the activity of the snake venoms, particularly those of the viperids, is probably more important in cases of human envenomation. In addition to an exogenous activation fed by proteolytic enzymes, most often serin-proteases, the snake venoms stimulate all the endogenous pathways but principally the plasminogen activators of tissue origin, notably urokinase. The activation of plasmin allows the hydrolysis of fibrin to take place and to a lesser degree that of fibrinogen as well.

The analysis of the products of fibrin degradation after a viperine envenomation shows a predominance of FDP_g, which are products of non-stabilized fibrin exclusively; in effect these FDP_g cannot come from a degradation of fibrinogen which has been used up rapidly from the beginning of the envenomation and is no longer available. Also the absence of FDP_n from the hydrolysis of stabilized fibrin is explained by the formation of an abnormal thrombus under the action of the thrombin-like enzymes present in the venoms of vipers and crotalids.

Certain chemical inhibitors amongst those that are used habitually in therapeutics appear to be effective on snake venoms *in vitro*. Their *in vivo* activity has not yet been subjected to systematical therapeutic research.

5.2.2.3 Inhibition of coagulation

Several proteinase inhibitors have been isolated from snake venoms. Their mode of action has not been elucidated entirely. These proteins are rarely specific: They act simultaneously on several coagulation factors. In any case their clinical importance is admittedly minor.

Inhibitors of factors IX and X have been isolated from the venoms of *Deinagkistrodon acutus* and *Echis leucogaster*. A thrombin inhibitor, bothrojaracin, has been extracted from the venom of *Bothrops jararaca*. All these inhibitors bind to the

Table VII
Principal fibrinogenases of snake venoms

Serpent	Common name	fibrinogenolysis	fibrinolysis
Agkistrodon contortrix	fibrolase	α > β	α > β
Agkistrodon piscivorus	β-fibrogenase	+ β	+
Bitis arietans	PAV protease	+ α / β	+ γ
Bothrops asper		+ α > β	
Bothrops atrox		+ α	+ α
Bothrops jararaca	jararafibrase	+ α > β	+
Bothrops moojeni		+ α > β	+ α
Calloselasma rhodostoma	α-fibrogenase	+ α	+
Calloselasma rhodostoma	kistomin	+ α > γ	+
Cerastes cerastes	cerastase	α > β	α > β
Crotalus atrox	atroxase	α > β	α = β
Crotalus atrox	catroxase	α > β	-
Crotalus atrox	β-fibrogenase	β > α	-
Crotalus basiliscus	basiliscusfibrase	α > β	α > β
Crotalus basiliscus	basilase	α = β	α > β
Crotalus molossus	M 4	+ α > β	+ α / β
Crotalus viridis	CVO protease V	+ α / β	+
Deinagkistrodon acutus		+ α	+ α
Echis carinatus	fibrinogenolysin	+ α > β	+ α > β
Gloydius blomhoffii		+ α / β	+ α / β
Gloydius halys	protenase L4	α	+
Gloydius halys	β-fibrogenase	α > β	α > β
Lachesis muta		+ α	+
Macrovipera lebetina	fibrinogenase	β + α	
Macrovipera lebetina	lebetase	α > β	α = β
Naja kaouthia		+ α	+ α
Naja naja		+ α	+ α
Naja nigricollis		+ α	+ α
Naja nivea		+ α	+ α
Ovophis okinavensis		+ α	
Philodryas olfersi		+ α	+ α
Protobothrops mucrosquamatus	α-fibrogenase	+ α	+
Protobothrops mucrosquamatus	β-fibrogenase		β > α
Trimeresurus gramineus	α-fibrogenase	+ α	+ α

corresponding factor while it is inactive and prevent them from combining with the corresponding activation complex. It is a classical competition phenomenon.

5.2.2.4 Hydrolysis of the phospholipids
The phospholipids play a fundamental role in coagulation by acting at different levels. They are an essential element in the architecture and organization of the cytoplasmic membrane, particularly of the cytoskeleton of the blood platelets. The platelet phospholipids release thromboxane A_2, a powerful vasoconstrictor, and the endothelial phospholipids generate arachidonic acid, the precursor of prostaglandins, which are important mediators of inflammation. Finally the tissue and serum phospholipids render an indispensable support to the initiation of several biochemical reactions, in particular the fixation of the "prothrombinase" complex.

Clinically the phospholipases act predominantly at the level of the membranes. However, the phospholipases A_2 of the venoms of elapids release lysophospholipids that can be of tissue or serum origin. Their tensio-active effect provokes the destruction of erythrocytes. The toxic effects appear to be secondary even if the haemolysis is severe.

5.3 Effects on the blood platelets

The snake venoms possess many substances that can act on the thrombocytes. The activation or inhibition of the blood platelets, sometimes observed in the same venom, have in principle no severe clinical expression.

5.3.1 Activation of platelet aggregation
Platelet aggregation is an indispensable mechanism for the formation of the haemostatic plug and it is enhanced by several types of proteins in snake venoms.

Most of the serin-proteases extracted from snake venoms stimulate the platelets. They can provoke simultaneously the aggregation of the platelets and their degranulation. Cerastobin, a serin-protease extracted from the venom of the North African viper *Cerastes vipera*, can activate the platelets while at the same time allowing the hydrolysis of fibrinogen. Thrombocytin extracted from the South American crotalid *Bothrops atrox* has a mode of action on the platelet membrane similar to that of thrombin, recognizing the same receptors. Cerastocytin from the venom of *Cerastes cerastes* has the same properties. However, cerastatin also from the venom of *Cerastes cerastes* binds to a different site from that recognized by thrombin. Crotalocytin from the venom of *Crotalus horridus* is more powerful than thrombocytin and probably has a different mode of action: It agglutinates the platelets in the absence of exogenous fibrinogen like natural thrombin. Like all the serin-proteases in snake venoms, those that induce platelet activation have a very variable sensitivity to the different natural thrombin inhibitors: antithrombin III, heparin and hirudin.

The agregoserpentins that are found in the venoms of many American and Asiatic crotalids do not have an enzymatic activity. Calcium is indispensable for their activity. Under their action the platelets do no longer react in the presence of collagen, one of the principal inductors of their activation. This suggests that the mode of action of the agregoserpentins is similar to that of collagen and like the latter is expressed by the aggregation of the platelets. However, the agregoserpentins are not inhibited by salycilates (aspirin and indometacin) with the exception of the one isolated from the venom of *Protobothrops mucrosquamatus*. Triwaglerin from the venom of *Tropidolaemus wagleri* activates the phospholipase C.

The thrombolectins stimulate the release of the cytoplasmic granules and their contents allow the platelets to aggregate. They would act by binding on a yet unidentified membrane receptor.

Finally the phospholipases A_2 hydrolyse the membrane phospholipids that release arachidonic acid, the inductor of platelet aggregation. However, some phospolipases provoke a platelet aggregation independently of the release of arachidonic acid, even in the presence of an inhibitor of the latter.

5.3.2 Inhibition of platelet activation

Many enzymes that hydrolyse the inductors of platelet activation are present in snake venoms; they have an indirect anti-aggregating action. Bothrojacaracin from the venom of *Bothrops jararaca* is a specific inhibitor that acts on four levels. It inhibits the binding to thrombomodulin, a membrane protein that activates protein C. It reduces the activation of protein C by α-thrombin. It prevents the association between α-thrombin and fibrinogen. Finally it enters in competition with heparin at two binding sites on the thrombocyte membrane. Bothrojaracin causes an inhibition of thrombin both when free and when bound to the blood clot.

Several phospholipases A_2 act like inhibitors of thrombocyte activation. The modes of action, as far as they are known, are variable. The alteration of the cytoskeleton and the increase in the intracellular concentration of cyclic adenosin monophosphate (AMPc), which assures the transmission of the hormonal signal into the interior of the cell, are the principal causes of the inhibition of platelet aggregation.

In the same way dendropeptin enhances the release from the endothelium of AMPc and of cyclic guanosine monophosphate (GMPc), which plays the same role as AMPc.

Although mambin from the venom of *Dendoaspis* has a structure of the type "neurotoxin with three fingers", it does compete for the fibrinogen-integrin bond and inhibits platelet aggregation.

The disintegrins enter into competition with the mediators of platelet activation at the level of the integrins, which are transmembrane proteins with receptors responsible for intercellular exchanges and for cyto-adherence. Thus they oppose the formation of the platelet plug by inhibiting platelet aggregation.

Certain metallo-proteases are active on the platelets; their yield appears to be proportional to their molecular mass. The addition of new protein domains potentiates their enzymatic activity. The metallo-proteases of group 3 which in one part of their molecule show a strong analogy with the disintegrins and those of group 4 which add a domain rich in cystine, block platelet aggregation.

6. Necrotizing Activity

Viperine envenomation combines a strong local inflammatory and necrotizing activity and this is grouped under the name viperine syndrome which is distinct from the complex haematological symptomatology which is important for the immediate vital prognosis. This severe syndrome causes some 100,000 amputations annually throughout the world.

6.1 The pathogenesis of the viperine syndrome

The local and regional process seen in a case of viperine envenomation is caused by four strongly synergic factors that are difficult to dissociate (Figs. 24 and 25).

6.1.1 Non-specific enzymatic activity

The enzymes in the venom of viperids have strong hydrolytic capabilities. The phospholipases A_2 have an effect on the membrane polysaccharides. The hyaluronidases hydrolyze the mucopolysaccharides of the connective tissue, which enhances the diffusion of the other components of the venom. The proteases attack the different structural tissues, muscle, bone or endothelium, as well as the functional proteins like certain coagulation and complement factors and diverse chemical mediators.

6.1.2 Inflammatory response

The action of phospholipases on the cell membranes produces in addition to structural destruction the release of arachidonic acid, which is the precursor of a number of strongly inflammatory substances, notably the leucotrienes that increase capillary permeability and the prostaglandins that cause vasodilation and potentiate bradykinin and the thromboxanes. The presence of plasmin sets in motion the kinin system, which could also be activated directly by the venom. Bradykinin is strongly vasodilatory which accentuates the hypotension and extravasation induced by the destruction of the endothelium and is expressed by the apparition or increase of edemas or rushes. In addition it causes pain. The stimulation of the complement system by the venom causes on one hand the formation of histamine which is also induced by plasmin and/or certain enzymes of the venom, and on the other hand the direct production of bradykinin. Histamine triggers the relaxation of the smooth fibers of the arterioles and the contraction of the efferent venules, which results in capillary stasis and furthers extravasation. Finally the activation of the cellular immune system causes the release of cytokines with multiple effects on inflammation and the defenses of the organism in general. Interleukin 6 (IL-6) appears early and stimulates the synthesis of inflammatory proteins. IL-8, which induces the adhesion of monocytes to the vascular endothe-

The Toxicology of the Venoms

Figure 24
Aetio-pathogenesis of necrosis

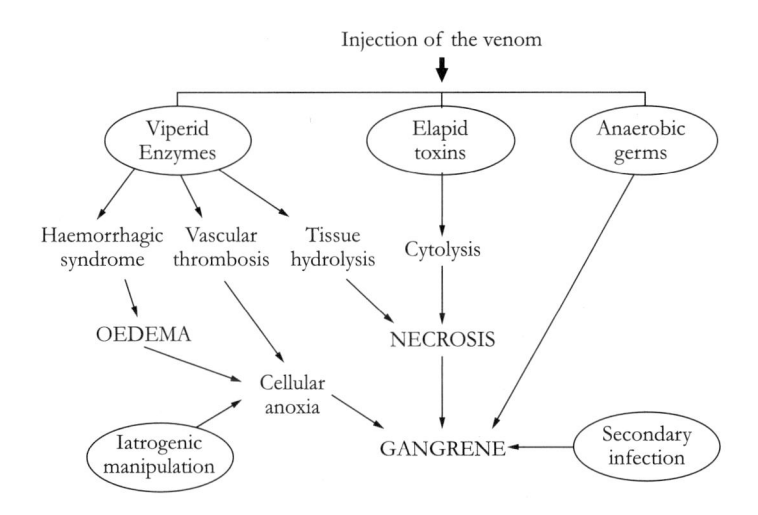

Figure 25
Mechanisms of inflammation

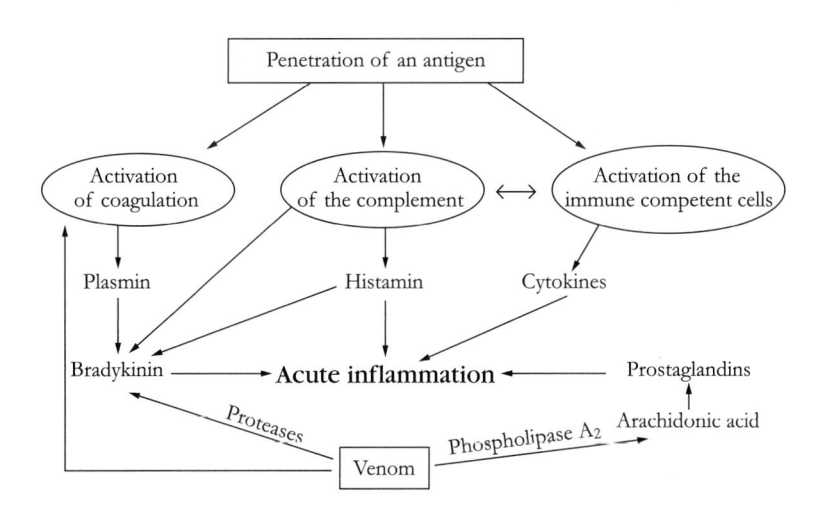

lium, appears much later. One does not observe the production of IL-1, another cytokinin which enhances the adhesion of the leucocytes to the endothelium and which increases the synthesis of prostaglandins, nor that of the tumor necrosis factor (TNF) that has similar properties. In the end the consumption of the coagulation factors under the direct action of the venom or linked to the destruction of the endothelium leads to a lasting inability of the blood to coagulate, which further amplifies the extravasation of blood. In this way many factors contribute to augment the inflammatory response and the edema that has a strong tendency to expand.

6.1.3 Infection-inducing factors

The third factor that intervenes is secondary infection. The oral cavity of snakes is strongly septic. Many bacteria have been found there, notably *Enterobacter* and *Pseudomonas*, which can be introduced during the bite. Also the bacteria present on the skin of the victim and on the instruments eventually used for treatment, can contaminate the wound. First aid application, particularly certain traditional practices such as scarification and plasters, but also surgical interventions, are an evident source of secondary infections.

6.1.4 Iatrogenic factors

The last factor is mechanical: therapeutic actions which aim to slow down the diffusion of the venom or to extract it, contribute in effect to arrest the blood circulation and cause an anoxia in the tissues. The tourniquet is the most frequently used first aid application in tropical countries. Sometimes it stays in place for several hours, even days. Local incisions, another very common practice, alter blood circulation and tissue perfusion and increase the contact surface between venom and cells.

One must well distinguish between necrosis due to the generally circumscribed tissue action of the venom and gangrene which is induced by tissue anoxia and secondary infections and which affects a larger region and generally is of iatrogenic origin.

6.1.5 Specific myotoxicity

The venoms of crotalids induce necroses of the skeletal muscles. These are both local, at the site of the injection of the venom, and systemic. In the venoms of

Crotalus spp. four groups of molecules have been identified which are responsible for muscle necroses: cardiotoxins, haemorrhagins, myotoxins and phospholipases A$_2$. In the *Bothrops* spp. it is mainly the phospholipases A$_2$.

The action of the cardiotoxins, also called cytotoxins, is non-specific: they hydrolyse the cell membrane.

The haemorrhagins destroy the basement membrane of the vascular endothelium and this provokes a muscle anoxia either because of lacking vascularization or because of the constriction caused by the edema.

The myotoxins enter into contact with the cell wall, bind to particular receptors and cause an alteration of the permeability for calcium, leading to a shrinking of the cell and a localized necrosis. This effect takes place progressively until the whole muscle fiber is destroyed. This phenomenon can be visualized perfectly by electron microscopy and manifests itself through a strong increase of the lactate-dehydrogenases, enzymes present in the cytoplasm of muscle cells.

The myotoxin of *Philodryas olfersii* hydrolyses the transverse striations of the skeletal muscles and causes the degeneration of the muscle.

The phospholipases A$_2$ alter the cytoplasmic membrane. One can compare their action with that of the β-neurotoxins, which are described further below.

6.2 Pathogenesis of the necroses caused by elapids

Certain elapids cause dry and limited necroses that develop into a localized mummification of the site of the bite. This necrosis can have two aetiologies.

The first one is specific. It is a muscular necrosis due to the phospholipase action of the β-neurotoxins of Australian and marine elapids and which are also found in certain viperids. A similar clinical, histological and biological symptomatology has been observed in envenomations by certain *Micrurus* spp., notably *M. nigrocinctus*, *M. alleni*, *M. frontalis* and *M. carinicauda*. It manifests itself by a myoglubinuria, a destruction of striated muscular fibres and a considerable rise of serum creatinine-phosphokinase (CPK). Its concentration is proportional to the amount of inoculated venom as well as to the microscopically determined myo-necrotizing effect. Thus the CPK level indicates the intensity of the muscle necrosis. The relative proportion of the different iso-enzymes allows also to identify the affected organ and to localize the site of the necrosis.

The other one is non-specific and due to the cytolysis which is provoked by the cytotoxins which are particularly abundant in the venoms of elapids. In the first stage of the necrosis caused by the venom of *Naja nigricollis* one can observe a vascular congestion, an edema and a degeneration of the basal layer of the skeletal muscle around the point of inoculation. The signs of transformation appear within 5 minutes. One hour after the injection of the venom an infiltration by a mixed population of cells is noted. Twenty-four hours after the injection of the venom the necrotic zone and its periphery are strongly colonized by polynucleated neutrophils and eosinophils, macrophages and mast cells. A depot of fibrin is seen in the blood vessels. After 18 days an epithelium and a basal layer of the muscle are in a stage of repair. The destroyed tissues leave behind a fibrous scar tissue that is complete after 28 days. The myo-necrosis is caused by one or several specific components of the venom, which is shown by the speed with which the lesions appear. However, it is probable that the lesions are partially caused (or enlarged) by hypoxaemia due to both the deposition of fibrin in the vessels and the edema.

The phospholipases of which the venoms of elapids are equally rich, can also explain the pain and the edema, though localized but present, which are seen in envenomations by *Naja nigricollis, Naja mossambica* and *Dendroaspis*.

The other factors, secondary infection and tourniquet, can very well add to the direct effects of the venom and lead to a more severe condition.

7. Toxicokinetics

After intravenous injection the venom is distributed within minutes throughout the body and then is slowly eliminated in a linear fashion. The half time of elimination of the venom is approximately 12 hours. Three days after the injection there remain less than 3.5% of the initial dose.

The intra-muscular injection is more representative of a natural bite and causes a more complex syndrome. Within 10 to 15 minutes the venom enters the circulation and its levels increase progressively, reaching a maximum within a few hours. Certain clinical studies suggest that this peak concentration is reached in less than half an hour, while experimental injections show that it occurs in approximately 2 hours. The diffusion of the venom in the body is slower than after using the intravenous route. It is probably governed by a law of mass diffusion and begins with a phase of very steep rise which after about 10 minutes becomes less steep then reaches a plateau before declining again progressively. The process of resorption stretches over 72 hours and this contributes to the maintaining of high plasma levels of the venom. The total bio-availability of the venom is about 65% of the injected quantity and does not vary with the dose that is administered.

Irrespective of the experimental conditions, notably the quantity of venom given (Fig. 26), the distribution of the venom is rapid (a few minutes to a few hours) while its elimination is slow (in the order of 3 to 4 days).

The distributed volume is higher than the volume in the plasma and this demonstrates the transfer of venom from the vascular to the cellular compartment. The fraction of non-modified venom excreted in the urine is constantly less than 5%.

Figure 26
Diffusion of the venom in the vascular compartment after intramuscular administration
(after Audebert *et al.*, 1994)

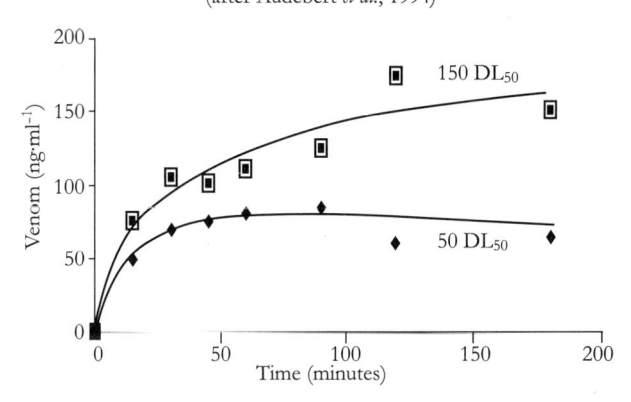

To start with the venom passes into the superficial tissues where it reaches a peak within 15 to 20 minutes for the venom of *Naja* for example. In the deep tissues, to which the toxin rich venoms of elapids diffuse preferentially, the concentration of venom is two to three times higher than in the blood (Fig. 27).

Figure 27
Diffusion of venom and antivenom in the anatomical compartments
(after Ismail *et al.*, 1996)

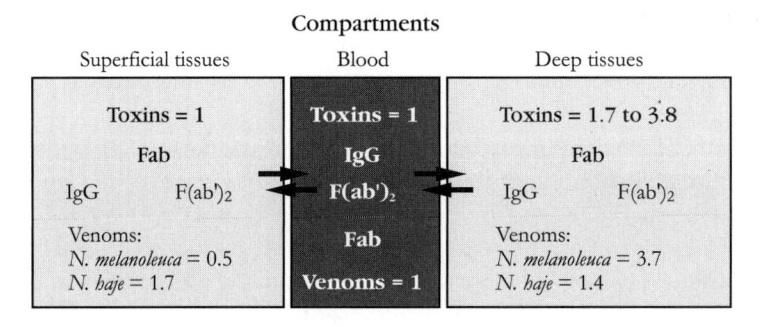

Compartments

Superficial tissues	Blood	Deep tissues
Toxins = 1 Fab IgG F(ab')$_2$ Venoms: *N. melanoleuca* = 0.5 *N. haje* = 1.7	Toxins = 1 IgG F(ab')$_2$ Fab Venoms = 1	Toxins = 1.7 to 3.8 Fab IgG F(ab')$_2$ Venoms: *N. melanoleuca* = 3.7 *N. haje* = 1.4

The maximum venom concentration in the deep tissues is reached within 1 to 4 hours depending on the species. This explains the appearance of clinical signs between 30 minutes and 1 hour and their progressive increase in 4 to 8 hours. The venoms of viperids appear to diffuse more slowly. They accumulate more in the liver and the kidneys. In the case of an intra-muscular injection a large proportion of the venom is retained around the point of injection.

After that, one observes the phenomenon of redistribution of the venom from the deep tissues to the central compartment for which it has however a lesser affinity. This redistribution is probably enhanced by the elimination of the venom following the law of mass distribution. To this phenomenon can be added a slow release of the venom, which has been confirmed by many experimental and clinical studies. This hypothesis could explain the re appearance of the venom in the circulation 7 to 10 days after the bite, observed by many clinicians after the treatment had been stopped. The precise site of retention of the venom in the body is not known. Certain specialists suggest that the venom is bound at the site of the bite from where it is released progressively. However, most think that it is kept in the lymphatic system, a hypothesis indirectly supported by the ineffectiveness of the instrumental aspiration of the venom at the point of the bite.

The venom is eliminated via the urinary and digestive route.

8. The Utilization of Venoms in Research and Therapeutics

The great efficacy of venoms and the diversity of their pharmacological properties have always fascinated naturalists. The ancients, above all Aristotle, have exploited the rich symptomatology seen after snake bite envenomations to include diverse snake products, venom but also blood and viscera, in the remedies used in those times. Theriac is an example of this and has been perpetuated until recent times.

The rationalism of the 18th century and then the experimentation of the 19th century and finally modern technology have allowed a development of the study of venoms and their utilization in biological research as well as in pharmacology.

The venoms of snakes are not the only ones to be utilized in research and in medicine. All the zoological groups have been involved: invertebrates (worms, sea urchins, cones [snails], insects, scorpions, ticks and spiders) and vertebrates (fish and even mammals) with the exception of birds.

8.1 Uses in basic research

In 1901 Richet discovered anaphylaxis while studying the venom of sea anemones, which have histamine-like effects that exacerbate a physiological reaction. A few years later Arthus working with snake venom defined hypersensitivity. A factor present in cobra venom activates the fraction C_3 of the complement, which is an essential factor of the inflammatory response. Secondarily the consumption of the C_3 leads to its depletion and indirectly induces a state of immune tolerance. The science of allergy was born.

At the beginning of the last century Calmette described the haemolysis caused by cobra venom. In 1911 Delezenne explained this phenomenon by demonstrating the effect of phospholipase on lecithin and the tensio-active effect of lysolecithin.

In 1949 Rocha e Silva isolated the bradykinins that are released during the penetration of crotalus venom into the body. The bradykinins are released spontaneously during the local inflammatory response. The viper venoms induce and amplify this natural phenomenon in the body.

The renin-angiotensin system has been explored with the help of the inhibitor of the conversion enzyme of angiotensin I to angiotensin II, which was extracted from the venom of *Bothrops jararaca*. Angiotensin II is a powerful vasoconstrictor. It increases the peripheral resistance and in so doing causes the arterial pressure to rise.

In 1957 Sasaki purified the neurotoxin of *Naja siamensis,* which reproduces entirely and exclusively the curare-like effect of the venom discovered a century earlier by Claude Bernard. With this tool it has been possible to purify the cholinergic receptor of the postsynaptic membrane and to describe precisely its molecular structure. The diverse but remarkably specific toxins obtained from venoms of elapids allowed the demonstration of the complexity of this receptor and the diversity of its active sites. The stimulation of each one of them explains the competitive inhibitory effects of certain substances that can be used in pharmacology and in therapeutics. Other toxins affect distinct receptors that are sensitive to L-glutamate, to the GABA system, to catecholamines or to other mediators of the central or peripheral nervous system. It is interesting to note that there is a close relation between the synaptic transmission and the growth of the neuron or the multiplication of the inter-neuronal connections.

The use of α-bungatoxin marked with radio-active iodine (I^{135}) has allowed one to understand the auto-immune etiology of myasthenia gravis, a disease which manifests itself by a severe insufficiency of muscular contractions leading to a complete motor paralysis. In fact the number of cholinergic receptors is considerably lowered as a consequence of their progressive destruction by the antibodies produced by the patient against them. Many pathologies linked to a dysfunction of the acetylcholine receptors (Parkinson's disease, Alzheimer's disease, depression, schizophrenia, neuralgias) can be studied with the help of the neurotoxins. The toxins that affect the potassium channel, like the dendrotoxins, have allowed one to demonstrate the progressive disappearance of the Kv1.1 receptors of the potassium channels in senescence and their early and significant reduction in Alzheimer's disease and this opens new therapeutic perspectives.

The nerve growth factor (NGF) discovered by chance in cobra venom enhances the rapid multiplication of peripheral nerve fibers. This peptide protects the neurons and orientates the connection of the neuronal fibers amongst each other. Used in cell cultures the NGF enhances the development of synapses and has allowed one to explain this process.

Watson and Crick used the nucleotidases of the venom of *Naja*, particularly the 5'nucleotidase for determining the structure of the DNA molecule. These enzymes have long been used to determine the base sequences of the different nucleic acids before the discovery of the restriction enzymes.

The amino-acid-oxidases of snake venoms are remarkably specific; they have been used to verify the isomerism of synthetic amino acids and to purify α-cetonic acids.

The phospholipases are used to study the physiology of the thrombocytes and of the pathological disorders that cause certain thrombo-embolic accidents. The very great specificity of venom phospholipases has allowed to demonstrate the anomalies of the phospholipids of the platelet walls of certain patients with a history of thrombo-embolic accidents. The level of arachidonic acid in particular is significantly different in these latter patients from that in persons without a history of thrombo-embolisms.

The phospholipases A_2 of snake venoms are used to solubilize the intra-membrane proteins and explore the functions of the different phospholipids that play a role in cell wall physiology.

Other enzymes involved in blood coagulation disorders have also been used to study this complex physiological phenomenon.

8.2 Uses in diagnostics

The lipid disturbances caused by certain metabolic or infectious disorders can be elucidated by reactions that use snake venoms. One of the first, Weil's reaction, was in common use at the beginning of the 20th century for the early diagnosis of syphilis. At the start of the treponema infection the blood cells of the patient are abnormally sensitive to the lytic factors of cobra venom.

In 1952 a pregnancy test was developed which exploited the sensitivity of horse erythrocytes to serum lysophospholipids, which are more abundant in pregnant women, after their release by cobra venom phospholipases. The fact that this property was non-specific soon became apparent as an unacceptable limiting factor.

The strong covalent bond between the α-neurotoxins and the cholinergic receptor is exploited for the titration of the level of antibodies produced against the receptors in patients suffering from myasthenia gravis. The cholinergic receptors coupled to α-neurotoxins are exposed to the serum of the patient. The antibodies against the receptors bind to the preparation. After the elimination of the excess reagents it is possible to titrate the quantity of receptors saturated by the antibodies and to measure the concentration of the latter. The disease expresses itself with symptoms identical to a cobra envenomation and its treatment uses an immunodepressant medication (corticoids, cyclophosphamide, azathioprin) and at the same time acetylcholinesterase blockers (neostigmin, edrophonium).

The enzymes involved in blood coagulation are widely used for the diagnosis of coagulation problems. Amongst them the thrombin-like ones simulate the effect of natural thrombin without reproducing it completely. They allow the application of coagulant treatments with local or general effects that are remarkably effective.

The thrombin-like enzymes from viperid venoms allow the study of the formation of fibrin, the thrombus formation time, due to the transformation of fibrinogen into fibrin, under certain pathological circumstances. The coagulation failure can be linked either to an insufficient level of fibrinogen, to the presence of a natural, therapeutic or toxic inhibitor, or to a congenital or acquired qualitative anomaly of the fibrinogen. Reptilase® time is used for detecting the presence of an inhibitor of fibrin formation. This haematological test consists of replacing thrombin used in the original thrombin time test by a thrombin-like enzyme which is not sensitive to the classical thrombin inhibitors, notably heparin, and this allows the titration in patients who have been treated with heparin. Reptilase® is made from batroxobin, a thrombin-like enzyme extracted from the venom of *Bothrops atrox* and *B. moojeni*.

Stypven® derived from the venom of *Daboia russelii*, is used as a reagent to produce prothrombinase directly without passing through the phases of tissue contact or platelet activation or antihaemophilia factors. Stypven® has properties which are analog to factor VII (proconvertin) and allows the activation of factor IX (antihaemophilia factor); this latter generates prothrombinase and constitutes the essential physiological way of the first step of coagulation. The activated factor IX obtained in this way differs morphologically from the factor IX that is activated naturally by the extrinsic route, which confirms that Stypven® acts directly on factor IX. In this way it is possible to localize a deficit in blood coagulation situated at the level of prothrombinase formation. Stypven® is also a direct activator of factor X in the absence of phospholipids as shown by the partial cleavage of the native molecule which produces an active Stuart factor that is different from the one formed naturally and which undergoes a second reducing cut. The venom of *D. russelii* is utilized for measuring the activity of factor X in patients with a coagulation deficit.

The plasma factor V can also be titrated with the help of various venoms, amongst them those of *D. russelii* and *Oxyuranus scutellatus*. A protease that activates prothrombin transforms prothrombin into thrombin in the presence of a complex containing factor V, phospholipids and calcium. This property is used by several tests, e.g. Chromozyme® TH which contains a chromogenic substrate. The hydrolysis of this substrate releases a pigment in a quantity proportional to the concentration of factor V in the plasma to be tested.

In the case of a vitamin K deficiency prothrombin is synthesized in an abnormal manner and does not present the carboxyglutamate groups that are necessary for its activation to thrombin. Ecarin obtained from the venoms of most of the *Echis* spp. can activate the incomplete fragment of prothrombin called acarboxyprothrombin that allows to confirm the existence of an abnormal prothrombin. Tests using a chromogenic substrate have been developed. There are many causes of vitamin K deficiencies: hepatic insufficiency, nutritional deficiencies and above all overdosing with anticoagulants using an antivitamin K. Ecarin is also used to supervise the coagulation time of patients which have been treated with hirudin, an anticoagulant extracted from the medicinal leech and given to patients which are allergic to heparin. Similarly, carinactivase 1 extracted from the venom of *E. carinatus*, which needs the presence of calcium and the integrity of the Gia domain of prothrombin, could be used to monitor oral anticoagulant treatments. The prothrombin activator present in the venom of *Pseudonaja textilis*, Textarin®, is active in the absence of phospholipids. This particular property is exploited in a test that allows the hydrolysis of prothrombin in patients suffering from a haemorrhagic lupus by producing antibodies against phospholipids.

Protac® extracted from the venom of *Agkistrodon contortrix* is used to artificially activate the protein C for carrying out its titration. Protein C is a glycoprotein produced by the liver and has a double anticoagulant effect. On one hand it inhibits the activated factors V (proaccelerin) and VIII (antihaemophilia factor A) and on the other hand it stimulates fibrinolysis. The Protac® test allows to verify the integrity of protein C. Its congenital and acquired failures are the cause of a large number of spontaneous thromboses, notably superficial thrombophlebitises and arterial thromboses.

Botrocetin from the venom of *Bothrops jararaca* is a powerful platelet aggregating agent and allows the diagnosis of several haemorrhagic diseases of genetic origin like von Willebrand's disease and the haemorrhagic thrombocytic dystrophy of Bernard and Soulier. The former is caused by a lack of the von Willebrand factor that is associated with factor VIII and necessary for the adhesion of the blood platelets. In contrast to the peptides usually used for titrating the von Willebrand factor, like ristocetin, which recognizes only its polymers of high molecular weight, botrocetin binds to all the conformations of the von Willebrand factor, whatever their structure in the plasma. This permits a more sensitive diagnosis of the different pathological forms of von Willebrand's disease. The disease of Bernard and Soulier is characterized by the absence of glycoprotein I on the thrombocyte wall and is expressed by an anomaly of their adhesion. Consequently the first step of haemostasis is strongly disturbed and these conditions are marked by a prolonged bleeding time.

The haemorrhagins of venoms of certain viperids have been proposed for the diagnosis of patients suffering from congenital or acquired capillary fragility.

8.3 Therapeutic uses

Treatments based on snake venoms have been part of folkloric or mystic repertoires for a long time. Not long ago the venom of *Vipera aspis* was used against clinical rabies (1835), the venoms of cobras and crotalids have been used for the treatment of epilepsy (1908) and viperid venoms for the control of haemorrhages (1935). With the exception of the latter that are still in use, all the other ones have been abandoned because of their inefficacy and danger.

Venoms also play a role in homeopathy in diverse indications.

Analgesic preparations containing venoms of *Naja* or of *Crotalus durissus terrificus* that were commercialized in the 1930s, have disappeared again in the 1960s. They had relatively important side effects, but some claimed anti-edematous, anti-inflammatory and even anti-tumor properties. Their main advantage was that they were not habit-forming like the morphines. While the pain-relieving properties of these heroic medicines were questioned by many practitioners, they were abandoned because of improvements of other treatments. If the anti-tumor properties are disputable, the synaptic block achieved by the neurotoxins, causing the immobilization of the joint, could explain its analgesic effect in rheumatism.

They were also prescribed as anti-convulsives.

Notexin is a pre-synaptic neurotoxin isolated from the Australian elapid *Notechis scutatus,* which destroys the nerve ends. It is used in local treatment to eliminate abnormal neurons found in congenital neuropathies and thus to allow the regeneration from healthy nerve cells.

The use of venoms as haemostatic agents goes back to the end of the 18th century, when Fontana in 1767 described the coagulant effects of viper venom. The complete, more or less diluted venoms of *Agkistrodon, Notechis, Naja* and *Bothrops* have been used in trying to treat spontaneous or accidental haemorrhages in haemophilia, trauma and diverse pathologies of haemostasis. A better identification of the venom components with coagulant activity and above all the development of purification techniques have led to the commercialization of several pharmaceutical specialties. At present Reptilase® and Bothropase® are sold as haemostatic agents for local treatment. Their importance in therapeutics has decreased since the development of substitution treatments and the use of natural thrombin of human or bovine origin. However, it could rise again because of the real risks inherent in this type of therapy, particularly the contamination with unconventional infectious agents (bovine spongiform encephalopathy), hepatitis viruses or that of human acquired immunodeficiency syndrome… The absence of cross immunogenicity between the different thrombin-like enzymes allows their use to be alternated.

Batroxobin (*Bothrops moojeni*) provokes the cleavage of fibrinogen into fibrinopeptide A and allows the release of fibrin 1, which stimulates the plasminogen activators of the vascular endothelium and activates protein C. The friability of the fibrin formed by batroxobin discards any risk of vascular thromboses. Also the fibrinogen levels are lowered by consumption and the anticoagulant activities of plasmin and of protein C will reduce blood viscosity and consequently also any thrombo-embolic risks. Bothrojararacin extracted from the venom of *B. jararaca* partially inhibits thrombin by blocking the receptors of fibrinogen and heparin while leaving certain catalytic properties, which give it an anti-thrombosis activity. Ancrod (*Calloselasma rhodostoma*) has for a long time been used as haemostatic under the name of Arvin® and today it is offered as anti-thrombotic under the name of Viprinex®. Several of these thrombin-like enzymes are presently objects of promising clinical trials in the United States, in China and notably in Brazil. A limiting factor of this kind of treatment during its prolonged administration could possibly be the development of antibodies to the enzyme that often has a high molecular weight. However, this risk is now limited by the large range of thrombin-like enzymes, which are hardly related immunologically to each other.

The control of thromboses benefits equally from the metallo-proteases, which have a fibrinolytic or fibrinogenolytic activity. Fibrolase extracted from the venom of *Agkistrodon contortrix* and the plasminogen activator isolated from the venom of *Trimeresurus stejnegeri*, TSV – PA, could be utilized as fibrinolytic agents in vascular embolisms. These two molecules can be produced by genetic engineering through recombination.

Thrombin is used in surgery as a biological suture to obtain the adhesion of two cellular surfaces. The catalytic action of thrombin on fibrinogen allows the *in situ* polymerization of fibrin with tissue proteins. The risks of contamination inherent in natural thrombin give preference to the thrombin-like enzymes of viperid venoms. Vivostat® (batroxobin) is used in surgery since 1995. Batroxobin is incubated with the patient's serum. The batroxobin is removed from the clot of fibrin 1 at an acid pH, and then placed into solution again with a neutral swab to be applied to the wound where it causes an immediate polymerization. This type of biological suture has the advantages of being diffuse, resistant, resorbable and clean.

The inhibition of cellular adhesion provoked by the disintegrins could be put to use under many circumstances, notably certain thrombo-embolic pathologies due to platelet aggregation. Echistatin is offered under the name of aggrastat (Tirofiban®) for the prevention of thromboses in angina pectoris.

Lebecetin, a disintegrin of the venom of *Macrovipera lebetina*, is the object of experimental research with a view to its use as anti-cancer drug, particularly against melanoma. Eriststatin, a disintegrin extracted from the desert viperid *Eristocophis macmahoni*, does not have any direct cytotoxic action on malignant melanoma metastases but induces an apoptosis, a premature self destruction of these cells by an intrinsic cytotoxic action. Contortrostatin extracted from the North American crotalid *Agkistrodon contortrix* inhibits the adhesion of many cancer cells to structural proteins, which prevents them from passing through the basement membranes and blocks their dispersion in the body. Several viperid disintegrins share this property and they could thus limit the metastatic process that is the cause of dissemination of cancer cells in the body.

From the venoms of *Bitis arietans* and *Calloselasma rhodostoma* it has been possible to purify factor VIII (antihaemophilia factor A), which can be given to haemophiliacs.

A substance of small molecular weight isolated from *Bothrops jararaca* blocks the degradation of bradykinin, which is a powerful vasodilaror. This molecule can also inhibit the enzyme which allows the conversion of angiotensin I to angiotensin II, a vasoconstricting hormone of the renin-angiotensin-aldosteron system. Aldosteron provokes a sodium retention and loss of potassium, factors which enhance arterial hypertension by increasing the extracellular volume. This substance thus acts on two levels: it causes on one hand a drop of the plasma level of angiotensin II and of aldosteron, both implicated in raising arterial pressure, and on the other hand a rise in the level of bradykinin. The drugs that have been synthesized on the model of this molecule (Captopril®, Lopril®, Captolane®) reduce the plasma volume and the peripheral resistance, which are the principal causes of arterial hypertension. In addition these drugs do not have a rebound effect, which is one of the serious side effects in this type of treatment.

Hyaluronidase is used for enhancing the diffusion of aqueous substances in edemas and haematomas. Its use in combination with drugs has been proposed to enhance their penetration to their site of action.

The cobra venom factor, or CVF, acts on two fractions of the complement, C_3 and C_5. The activation of C_3 is generally more important and leads to its depletion as consequence of a higher consumption. This property is experimentally exploited to induce a state of tolerance in cases of transplantation of heterologous cells. The administration of CVF twenty-four hours after the injection of myoblasts, indifferentiated muscle cells, enables them to survive and to colonize the muscle suffering from Duchenne's muscular dystrophy. In addition xenograph transplants, particularly in pigs, have been significantly prolonged in animals treated beforehand with CVF.

CHAPTER 4

Antidotes and Immunotherapy

Envenomations can be treated by intuitively following two complementary routes: the symptomatic approach that aims to correct the pathological effects caused by the venom and the inhibition of the active principle after its penetration into the body. Traditional and modern medicine both pursue the two objectives. Combining the empirical knowledge of the former with the experimental deductions of the latter could optimize the treatment of snakebites.

1. Traditional Medicine

The choice of plants used by traditional healers is based on a range of reasons some of which may appear to us far removed from Cartesian logic. However, the role played by empirical knowledge cannot be neglected and the results of centuries of medical observations are not without interest. Progressively we shall reach, via the clinical experience of practitioners, an experimental reasoning, which will correspond better with our rationalism.

1.1 Physiognomy

The evocative aspect of certain plants has sparked the idea of using them to treat a condition, which appeared to be linked directly to the image presented by the plant. In this way Paracelsus developed the "theory of signatures" according to which the form, the colour and even the taste express the therapeutic properties of each plant. *Rauwolfia serpentina*, Apocynaceae, of which is known that it has no antidotal effect, but well a tranquilizing effect, has roots which look perfectly like snakes, as indicated by its name.

1.2. Folklore

Old legends could also have suggested to the healer the virtues revealed by a divinity or a supernatural being more knowledgeable than we are. *Ophiorrhiza mungos* (Rubiaceae) would, according to Hindoo legend, be eaten by mongooses when they were bitten in a fight with a king cobra. No anti-venom effect of this shrub has been found to this day. The flower of *Feretia apodanthera* (Rubiaceae) would be an appreciated food of snakes: therefore it is used, crushed and swallowed with water, by a bitten person to prevent the envenomation.

1.3. The repellent effect

With the repellent effect we enter the domain of observation. Nobody can say with certainty by what means some healers discovered the repellent properties of plants. They might have observed the absence of snakes in certain areas with a strong specific plant population. Or they could have tested patiently the plants that are supposed to allow snake charmers to manipulate their charges more safely. However, experimental results often confirm these repellent properties.

1.4 The symptomatic effect

Without doubt the pharmacological action of certain plants gives relieve to patients presenting disorders as a consequence of the bite of a venomous animal. The treatment aims at combating the symptoms seen in an envenomation; evidently there is an interest in moderate or severe cases of envenomation. The reputation of a plant can be illusory or arbitrary in the light of criteria named above. It can also have derived from a verified effect after fortuitous discovery or after deductions based on clinical observations.

The symptomatic effect can either re-enforce an antidotal effect of be confused with it.

The principal reason for the use of the therapeutic preparation is to reduce the clinical problems seen during the envenomation. It will be directed to more or less intricate symptoms with variable importance for the patient, depending on the biting snake, the susceptibility of the victim and the therapeutic interventions that have taken place after the bite. Today the range of symptoms can be explained by a better knowledge of the composition of the venoms and the toxico-kinetics of their components, but the traditional pharmacopoeia follows a more clinical route, which isolates the symptoms and considers them independently of each other. These symptoms can be placed into two groups, respectively for the viperids and the elapids (Table VIII).

Table VIII
Comparison of symptoms of envenomations by viperids and elapids

Symptoms	Viperidae	Elapidae
Pain	frequent and severe	generally anaesthesia and paraesthesias
inflammation	frequent and severe	rare
digestive	sometimes	frequent and severe
œdema	extensive	rare and light
paralyses	no	yes
respiratory	no	yes
haemorrhages	abundant	no
shock / coma	yes	yes
necrosis	frequent and extensive	rare and limited

Many plants have properties that can be used for treating these clinical problems. Often they are administered together with recognized anti-venom drugs, from which they differ, though, by a more widespread use, often being available to the public. An exhaustive list of plants that can be used within this framework is presented in Annex 2. For some of these the symptomatic effect has not been confirmed and is in need of appropriate experimentation. Also there are sometimes side effects that can cause complications or accentuate the toxicity of the venom (Table IX).

1.4.1 The analgesic effect

This property is often confounded with an anti-inflammatory effect. The effect of the venom of viperids is particularly painful and sometimes needs a vigorous

Table IX
Desirable symptomatic properties for the use of anti-venomous plants

Symptoms	Desired activity	Side effects
Pain	analgesic, anti-inflammatory	haemorrhages
Œdema	anti-inflammatory	haemorrhages
Necrosis	anti-inflammatory	haemorrhages
Fear	sedative	Lowering of arterial pressure
Infection	antiseptic	
Haemorrhages	haemostatic	
Respiratory problems	respiratory stimulant	
Digestive problems	antispasmodic	respiratory depression
shock, low arterial pressure	cardiotonic circulatory stimulant	toxic
Fever	diuretic anti-inflammatory	Lowering of arterial pressure, haemorrhages
Paralysis	motor stimulation	toxic

analgesic treatment. The poppy, *Papaver somniferum*, and even certain Solanaceae (*Dartura* sp. *Nicotiana tabacum*) are currently prescribed by some traditional healers. Others make use of powerful sedatives like *Rauwolfia* sp., from which an antihypertensive, reserpine, and raubasine, an agonist of the benzodiazepines, have been extracted, or *Valeriana officinalis* from which the antispasmodic and febrifuge valerian is derived.

1.4.2 Anti-inflammatory action

The anti-inflammatory effect is currently used more widely. It has been observed with the flavonoids, the tri-terpenes, the sterols other than the corticoids, and the saponins. The flavonoids for example inhibit the enzymatic reactions that allow the synthesis of prostaglandins from arachidonic acid. Aristolochid acid is extracted from many plant species of the genus *Aristolochia*, which occur widely

Antidotes and Immunotherapy

around the world. This acid is a competitive antagonist of most of the phospholipases A_2 of snake venom, against which it has been tested. Cepharantin, an alkaloid extracted from *Cepharanthus cephalantha* (Menispermaceae) inhibits the phospholipases *in vitro*. The aqueous extract of *Diodia scandens* has an antihistamine and antiserotonine action. The antihistamines are essential humoral mediators of inflammation while the role of serotonine in inflammation is less evident. Velutinol A, a steroid obtained from the rhizome of *Mandevilla velutina* (Apocynaceae), strongly reduces the inflammation caused by the phospholipases,

1.4.3 Local action: anti-edematous, antiseptic and anti-necrosis

These symptoms are often poorly separated by traditional healers who confound or associate with each other pain, inflammation, edema, gangrene and necrosis and look for a general wound treatment. In any case, certain plants have more direct anti-edematous or antiseptic properties. It could be interesting to distinguish them from the anti-inflammatory plants, as they have a systemic mode of action, which could be of great interest even outside of envenomations.

1.4.4 Action on haemostasis

Several properties of the venoms are involved and it is difficult to determine the mode of action of those plants that show a real efficacy.

A local haemostatic effect that assures the protection of the vascular walls allows to compensate for the action of the haemorragins which destroy the vascular endothelium. The latter cause an extravasation of the blood responsible for the persistent bleeding at the site of the bite, at old scars and even on healthy endothelium, which results in a purpura or sero-haemorrhagic rashes. Certain plants have a haemorrhagic action by activating the coagulation system or by their proteolytic ability, dissolving the blood clot (papaya, pineapple for example); they will oppose the first stage of disseminated thromboses induced by the action of the venom. This proteolytic action, sometimes similar to that of physiological plasmine, can prevent the later complications of viperine envenomations.

These properties must be distinguished from the direct inhibition of the enzymes of the venom, which activate blood coagulation. This constitutes a specific antidote effect, as will be seen later.

1.4.5 Indirect supporting action: diuretic, cardio-tonic

Enhancing diuresis is a major worry of the traditional healer, who wishes to rapidly eliminate the toxin. Many plants have recognized diuretic properties and are widely employed in febrile conditions or intoxications to accelerate recovery.

Also in the presence of a somnolent or comatose state the traditional healer can administer a restorative, which quite often is a cardio-tonic.

1.5. Antidote effect

This effect can be confirmed through rigorous experimental study only. The antidote can act against the venom in two manners. The systemic antidote acts on a function that has been disturbed by the venom and act through competition or pharmacological antagonism. It protects or re-establishes the physiological function targeted by the venom. The specific antidote inhibits at the molecular level the toxic component of the venom. It directly opposes the venom itself.

1.5.1 Systemic antidotes

They concern a physiological function targeted by the venom. Three modes of action can be described: competition, antagonism and immune stimulation.

1.5.1.1 Competition
Competition leads to the substitution of the venom by the antidote at the level of the effector site. Depending on the toxicity of the competitor itself and its capacity to activate or inhibit the corresponding function, the antidote would be more or less effective. The substitution can be specific if the same effector site is touched or it could be paraspecific (or crossed) if a neighbour site is concerned but modifying the effector site sufficiently to prevent the toxic effect.

This competition could be linked to a molecular similarity as in the case of a peptide isolated from the bark of *Schumanniophyton magnificum* and the cardiotoxins of *Naja*. This peptide has a molecular weight of 6 kDa and an amino acid composition close to that of the cytotoxins of elapids. The inhibiting action of the peptide on the cardiotoxins is dose dependent, suggesting a competition at the level of the effector site of the cell wall. Atropine, the alkaloid of the belladonna and several other Solanaceae, is a competitor of the neurotoxins present

Antidotes and Immunotherapy

in the venoms of *Dendroaspis* (mamba), which bind selectively to the muscarin receptors of acetylcholin.

An extract of the root of *Securidaca longepedunculata*, a shrub of the African savannahs, apparently binds close to the cholinergic receptor of the post-synaptic membrane (Fig. 28) and thus prevents by steric hindrance the neurotoxin of *Naja* from attaching itself and inducing a paralysis through a neuromuscular block.

Figure 28
Mode of action of *Securidaca*

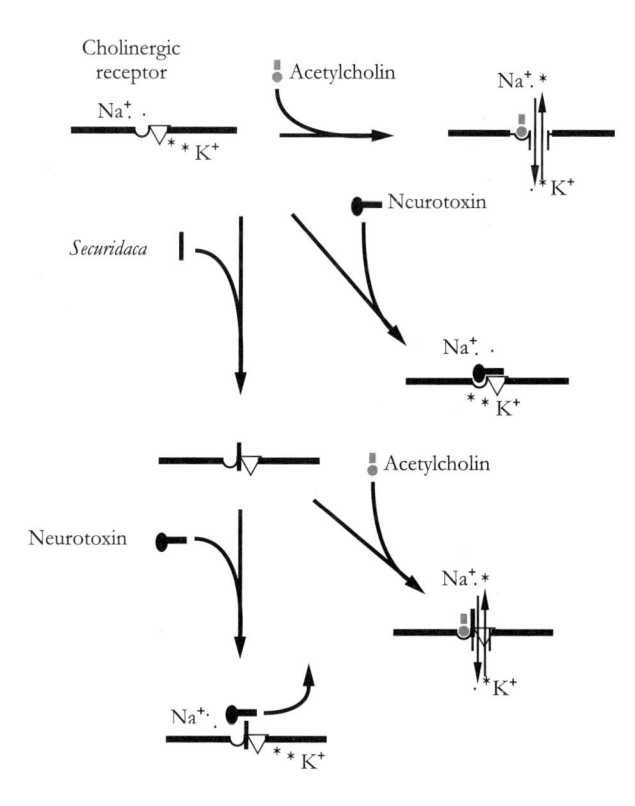

A similar action could explain the inactivation of the neurotoxin of *Naja siamensis* by an extract of *Cucurma* sp. The antagonism of this extract acts as well on a nerve-muscle preparation as on a mouse *in vivo* when the venom and the antidote are given by the same route. However, the extract of *Cucurma* does not inactivate d-tubocurarin.

Strophantus gratus, from which ouabain is extracted, inhibits the potassium pump and stimulates the calcium pump (positive inotropic and negative chronotropic effect), two systems which are much affected by the neurotoxins of snakes.

1.5.1.2 Antagonism

The antagonism depends on an opposite effect of antidote and venom on the same system. The venom triggers a reaction of the body, which is contested directly by the antidote.

The anti-cholinesterase effect of *Physostigma venenosum* (or Calabar bean) and of *Tabernanthe iboga* is similar to that of neostigmine. This latter inhibits the hydrolysis of acetylcholine and thereby maintains the mediator on its receptor, which prevents the neurotoxin from binding and allows the passage of the nerve stimulus.

Mandevilla velutina inhibits bradykinin, which plays an essential role in inflammation.

The alcohol extract of *Diodia scandens* (Rubiaceae), a plant that is widely used against snake bites in the Nigerian savannah, is inactive in normal plasma but inhibits the coagulation induced by the venom of *Echis ocellatus* when the latter is added to fresh plasma. The extract appears to oppose the effect of one of the coagulant factors present in the venom of *E. ocellatus*.

The roots of *Hemidesmus indicus* and to a lesser degree those of *Pluchea indica*, two Indian medicinal plants, inhibit at the same time the coagulant and anticoagulant factors of the venom of *Daboia russelii* and reduce significantly the toxicity of its venom.

Many plants have an effect on platelet aggregation as caused by the venoms of viperids. Thus *Capsicum frutescens*, chilli pepper, is anti-aggregating and *Aloe vera*, the aloe, is aggregating.

1.5.1.3 Immune stimulation

This property leads to the rapid elimination of the toxin by phagocytosis and the destruction of antigens before or after complex formation of the complement. The antidote can also control the proliferation of lymphocytes in response to the inflammatory stimulus and the production of lymphokines. A clear nonspecific stimulation of the immune response is seen after lymphocytes are placed in contact *in vitro* with *Boerhavia diffusa*. The increase in efficacy of the immunotherapy trough the treatment with plants is attributed to an improvement of the presentation of the toxic antigen to the antibodies. This has been confirmed with extracts of *Cucurma longa*, *Hemidesmus indicus*, *Dioda scandens* and *Okoubaka aubrevillei*.

1.5.2 Specific antidotes

They act directly on the venom and inhibit its activity. This inhibition can result from a molecular deformation, a direct chemical action on the venom or a major modification of the chemical environment. In the first two cases this efficacy depends directly on the probability of the two molecules meeting each other, resulting from the speed of their diffusion and their respective concentrations. The specific immunoglobulin perfectly fulfils this function when it binds to the antigen (enzyme or toxin of the venom), enhancing its precipitation and/or its elimination through the kidneys or the immune system. Certain plants have a specific inhibitory activity against the toxic enzymes of venoms. Plant extracts with an anti-phospholipase action have thus been identified.

The mode of action of punarnavosid, an anti-fibrinolytic extract of *Boerhavia diffusa*, is due to a direct inhibition of the fibrogenases of the venom, although the mechanism still has to be elucidated. Also the methanol extract of a South American cedar (*Guazuma ulmifolia*) inactivates the thrombin-like enzyme of the venom of *Bothrops asper* and significantly reduces its toxicity.

The ethanol extract of the roots of the Taiwanese plant *Aristolochia radia* inhibits the haemorrhagic activity of *Trimeresurus gramineus*, *Protobothrops mucrosquamatus* and *Deinagkistrodon acutus* but is without effect on the venoms of *Naja atra* and *Bungarus multicinctus*. The mode of action is still unknown.

1.5.2.1 Inhibition by altering the shape of the molecule

The shape of the molecule can be altered by the occupation of the active site of the toxic protein or of a nearby site, which modifies the conformation of the active site.

1.5.2.2 Inhibition by chemical modification

Some authors have described an oxidation or a reduction of the enzyme involved in one of the toxic actions of the venom. Also a total or partial hydrolysis of the toxic compound can be observed. This has been seen with schumanniofosid extracted from *Schumanniophyton magnificum*, or with the trypsine inhibitor extracted from the soya bean (*Soja hispida*).

1.5.2.3 Modification of the chemical environment

The chelation of a co-enzyme (notably calcium) or the mobilization of the substrate (fibrinogen for example) by the antidote with or without its transformation, can limit the activity of the enzymes of the toxin and consequently its toxicity.

1.6 Mechanical treatment

1.6.1 Tourniquet

This treatment was recommended by Galen in the 2nd century AC. Several experimenters in the 17th and 18th centuries, notably Francesco Redi and Felice Fontana, who were the first to study the action of viper venom in animals, confirmed these observations. More recently it was shown that the tourniquet does not prevent the death of the animal when the dose of the injected venom surpasses a certain threshold; however, all the authors agree that the tourniquet delays the onset of symptoms and that these are severely aggravated when the tourniquet is removed. Also the many and severe complications caused by leaving the tourniquet in place for too long a time are known: gangrene, which necessitates a large amputation, and even fatal septic shock. From this can be concluded that the tourniquet, although it presents the advantage of slowing down the diffusion of the toxin, bears two important risks: On one hand it concentrates the venom which will suddenly be released into the blood stream when the tourniquet is lifted and on the other hand it causes an anoxia in the distal tissues which is a source of major complications independently of the effects of the venom itself.

1.6.2 Aspiration

The extraction of the venom has been recommended ever since the envenomation has been associated with the penetration of an exogenic substance. The rationality of this method – the physical elimination of the venom through the point

Antidotes and Immunotherapy

of penetration – is evident. However, its efficacy has still not been demonstrated in spite of many attempts.

Sucking the wound, a romantic and probably inefficient method, can be dangerous to the operator, more because of his contact with the secretions of the victim, notably his blood, than by a hypothetical penetration of the venom into the mucosa or an oral wound of the helper.

The black stone, a fragment of bone toasted or burnt by craftsmen, and the various systems of instrumental aspiration of the venom (Aspivenin®, Extractor®, Venom-Ex®) belong into a category that is very close to traditional therapy. Originating in India the black stone arrived in Europe around 1650. It was mentioned for the first time in a 1656 document popularizing Far Eastern customs. According to this document both Chinese musk and the black stone would have come from the head of poisonous snakes. According to the then recommendations the stone is placed on the wound. There it adheres strongly while extracting the venom from it and releases itself spontaneously when the poison is absorbed. Subsequently it is regenerated by boiling it in milk and thus can be used indefinitely. The same recipe still exists today. The remedy has a strong symbolism. Its legendary origins, its jealously guarded production method, the "proofs" of its efficacy, its regeneration in milk, the traditional antidote, which in its virginal whiteness contrasts the black of evil, all contribute to its reputation. The black stone still is widely traded in developing countries.

The systems of instrumental aspiration have been subjected to experimental studies, which have shown their low efficacy. The venom diffuses in the body very quickly and the local effect of aspiration is too limited to render a sufficient yield. The amount of retrieved venom is negligible and the course of the envenomation is not altered significantly between treated and the control group. Also some authors draw the attention to the fact that the aspiration with the help of an apparatus using strong negative pressure enhances the development of necrotic lesions in the place where the suction apparatus has been applied. Therefore this method carries the risk of losing precious time and of causing an aggravation of the lesions, particularly when it is accompanied by the removal of debris or an incision of the skin.

1.6.3 Chemical denaturation

The local application of chemical substances is also practiced frequently. Various products are used and quite often their use lacks a scientific foundation. The various acids, bases, detergents and biological liquids (notably bile and urine) are founded more on irrational criteria than on rigorous experimentation. To inhibit the venom still present at the site of the bite is an interesting intervention if it is carried out early. Some enzyme inhibitors can play this role efficiently. However, one has to avoid aggressive substances that carry the risk of aggravating the lesions.

The use of disinfectants does not reduce the toxicity of the venom, but it helps to prevent secondary infections which play an important role in local complications.

1.6.4 Physical denaturation

Heat in the form of a flame or cautery provokes severe burns. Irrespective of the possible effect of heat on the venom, the burns cause severe lesions that are out of proportion with the improbable efficacy of this method. The electric shock of South American Indian origin is controversial. This treatment appears to be based on an Amazonian legend according to which the Indians treat the envenomations by electrocution with the help of an electric fish. Even if this curious practice should be real, it would not prove that it was efficient. It is painful and dangerous. The pseudo-scientific justifications – ionization of the proteins of the venom, stimulation of the defenses of the body or vascular electro-spasm that slows down the diffusion of the venom – are ridiculous in view of the total inefficiency proven in many clinical and experimental studies.

The rapid diffusion of the venom limits in most cases the applicability of these methods, which often have severe secondary effects.

2. Principles and Mechanisms of Immunotherapy

The therapy with antivenin is now more than a century old. After its discovery its therapeutic use has rapidly developed since the end of the 19th century.

Antidotes and Immunotherapy

Sewall immunized a pigeon in 1887 against the venom of *Sistrurus catenatus* by repeated injections of glycerinated venom. Even if the experimental protocol was not precise and not standardized, the pigeon could survive up to six times the lethal dose of the venom. Roux and Yersin demonstrated in 1888 that the blood of an animal immunized against diphtheria toxin protects a naïve animal against diphtheria. In 1890 Behring and Kitasato confirmed clinically the passive transfer of immunity against diphtheria and tetanus and by this fact invented serum therapy. Based on this discovery Chauveau proposed to apply to toxic cellular secretions the attenuation procedures used for microbial secretions. Under the influence and with the collaboration of Bertrand, Phisalix proved in 1894 the antitoxin property of blood of animals vaccinated against viper venom using a heat-attenuated venom. In the same year Calmette, who had worked on cobra venom in Saigon and later at the Institut Pasteur in Paris, studied three immunization protocols and observed like Phisalix and Bertrand that the serum of vaccinated animals also had a therapeutic effect. Calmette was the first to prepare an antivenom serum for medical use against the bite of Indian cobras and became the veritable promoter of the antivenin therapy which left the first results of Phisalix and Bertrand and even more so those of Sewall somewhat in oblivion. After Calmette who commercialized the antivenin against cobra venom in Saigon in 1896, many physicians developed antivenins in their countries or in the colonies belonging to their countries. Successively McFarland in 1899 in Philadelphia in the United States, Brazil in 1901 in Sao Paulo in Brazil, FitzSimons in 1903 in Johannesburg in South Africa, Lamb in Bombay and Semple in Kausali in India in 1904, Lister 1905 in London in the United Kingdom, Tidswell in 1906 in Sydney in Australia, Ishizaka in 1907 in Tokyo in Japan, all produced antivenom sera based on the protocols of Calmette.

Immunological progress rapidly led to anti-diphtheria and anti-tetanus vaccination: the diseases are caused by well-identified toxins which could be modified into non-toxic anatoxins which still have the same antigenic properties as the toxin. However, snake toxins, rich in varied substances, have up to now not lent themselves to human vaccination, even though interesting attempts exist. Antivenins still are the only specific treatment of envenomations: thus they are amongst the small number of centennial medications still in use. Although the mode of preparation of antivenins has not much changed since their discovery, the purification procedures and the mode of utilization have been modified profoundly and this has lead to the abandonment of the term serum therapy in favor of that of immunotherapy or antivenom therapy.

2.1 Production of antivenin

2.1.1 Immunization of the animal

The injection of raw venom is often poorly tolerated by the receiving animal. This has led to the preparation of an anatoxin by detoxifying the venom while conserving its immunological properties. Since the preparation of the first antivenins many techniques have been proposed for the detoxification of the venom. The most common ones are the combinations with aldehydes: formalin proposed by Ramon since 1924 or glutaraldehyde. Often an adjuvant is added to the detoxified or un-detoxified venom. The precise role of the adjuvant has not been elucidated, but it is thought that it slows down the resorption of the venom and strongly stimulates the immune reaction. Most commonly used are Freund's adjuvant, bentonite, aluminium hydroxide and sodium alginate.

The immunization protocol depends on the toxicity and the immunogenicity of the venom, the species of animal to be immunized and the quality of the response of the hyper-immunized animal. The quantity of injected venom has to be adjusted continuously by repeated controls to obtain a sufficient antibody titer. Ten to fifty injections spread over a period of three to fifteen months are necessary for a good immunization. When a suitable antibody titer has been reached the animal is bled onto an anticoagulant (sodium citrate) and purified, treated and then conditioned for its sale in liquid form or more rarely lyophilized. The animal of choice for the immunization is the horse, but other species can also be used.

The antivenins can be monovalent, when only one venom is used, or polyvalent, when the immunized animal receives a mixture of venoms from several species. The choice of one or the other type depends on many considerations. Principally a monovalent antivenin is more effective in the treatment of an envenomation by the corresponding species, but this is not the absolute rule. A toxic antigen that is present in small quantity in the venom of *Dendroaspis angusticeps* does not immunize the horses while a closely related antigen abundant in *D. jamesoni* protects the horses against the antigen of *D. angusticeps*. Consequently the polyvalent *angusticeps-jamesoni* serum is more effective against the venom of *D. angusticeps* than the

Antidotes and Immunotherapy

monovalent *D. angusticeps* one. It appears that the related antigens are acting synergistically to lead to a better immune response. Lists of producers of antivenins are drawn up regularly (cf. Appendix 3).

2.1.2 Purification of the antivenin and quality control

The antivenins, which used to be utilized in their untreated form, are now purified in successive stages to obtain a stronger efficacy and a better tolerance (Table X).

Table X : Preparation of anti-venom seraß

1. **Immunisation of the animal (horse)**
   - detoxification of the venom
   - utilization of an adjuvant
   - repeated injections (increasing doses)
   - collection of serum and verification of the protective activity

2. **Concentration of the protective fraction**
   - elimination of cellular elements by centrifugation
   - harvesting the γ-globulins with ammonium sulfate or the fixation of the β- and α-globulins and of thealbumin by ion exchange chromatography
   - elimination of the insoluble fraction by centrifugation
   - enzymatic hydrolysis by pepsin or papain to detach the Fc fracton of the IgG
   - coagulation of the Fc and inactivation of the complement at 57°C or separation of the Fc fragment by ultrafiltration through a membrane
   - elimination of the denatured proteins by centrifugation, filtration or chromatography
   - viral inactivation by heating in the liquid phase
   - harvesting γ-globulines with ammonium sulfate at a higher concentration
   - dialysis
   - elimination of fats by adsorption on aluminium hydroxide
   - antimicrobial sterilization by microfiltration through a millipore membrane

3. **Controls and standardization tests**
   - titration (neutralizing activity)
   - stability
   - specificity (precipitating activity)
   - quality tests (bacteriology – absence of pyrogens)

After the elimination of the cellular elements by centrifugation, the non-immune proteins, particularly albumin, are also removed after their precipitation with ammonium sulfate by a further centrifugation. Some producers carry out a chromatography for separating the immunoglobulins from the α-and β-globulins that have no immunological properties.

The albumin and those globulins that do not take part in the immune reaction can also be precipitated by caprylic acid. This technique appears to give a better yield than ammonium sulfate and to facilitate the purification of the IgGs or their antibody fractions. However, it has not been used at large scale because of the difficulty to standardize production.

Then the γ-globulins are subjected to a controlled proteolysis (Fig. 29).

Figure 29
Fragmentation of the therapeutic immunoglobulins

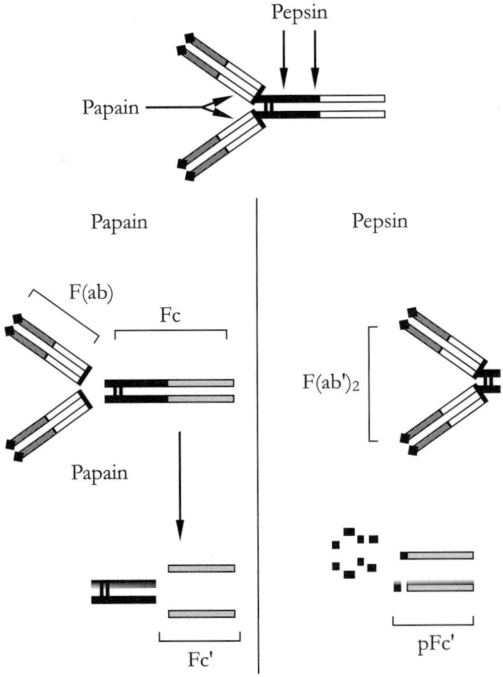

Antidotes and Immunotherapy

The immunoglobulin G (IgG) is a voluminous protein of a molar mass of 150 kDa, which is composed of two thermostable fragments Fab which carry the immunological function (Fab = fragment antigen binding) and a thermolabile fragment Fc which reacts with the complement and allows the precipitation of the complex formed by the antigen and the antibody and thereafter its elimination from the body. A digestion by pepsin separates the Fc fragment from the $F(ab')_2$ fragments, which have an average molar mass of 90 kDa and carry two antigen binding sites like the immunoglobulin. A papain treatment separates the Fc fragment from the individualized Fab fragments of an average molar mass of 50 kDa, which carry only one antigen binding site. These characteristics give the fragments $F(ab')_2$ and Fab biological properties which differ from those of the original IgG. The fragment Fc is separated from the specific fragments Fab and $F(ab')_2$ by ultrafiltration through a membrane using their difference in molecular mass.

A further purification chromatography is added by some producers.

Sterility can be achieved through heating against viruses and by filtration through a micropore membrane against bacteria.

The precipitation with caprylic acid allows one to obtain after an appropriate enzymatic digestion purified immunoglobulin fragments. It appears that this method avoids the contamination of the $F(ab')_2$ by IgG or other immunologially inactive globulin fragments. However, the Fab obtained after fractioning by caprylic acid would be of lower quality. The preparation, though, which is purified in this manner, lends itself to parenteral administration.

Before the final conditioning the product undergoes at each of the stages of purification various controls that allow:

- to verify its composition by physico-chemical tests;

- to demonstrate its innocuity by inoculating appropriate culture media, by immunochemical tests for purity, by toxicological tests, the inoculation of animals to confirm the absence of pyrogens;

- to guarantee its efficacy by immunological tests measuring the protective efficacy.

While the first two categories of tests are well codified, it is not entirely the same for the control of the protective power. A first immunological verification can be based on immunoprecipitation tests, by immuno-electrophoresis in a gel or by Western blot. The immunoglobulin fragments must react with all the protein fractions of the venom, which were separated by electrophoresis. In this way the different immune sera from different animals or from different production sources can be compared in this way, or cross reactions with other venoms can be demonstrated. It is even possible to evaluate quantitatively the precipitating antibodies of the prepared serum. However, a precipitating reaction does not necessarily imply a functional neutralization: the precipitating epitopes can be different from the neutralizing ones. Also some powerful immunogens may have no or only low toxicity and certain toxic proteins may be only weakly antigenic. The immunolocial controls do therefore not suffice.

The true standardization of the antivenins implies a verification of their neutralizing power, of their specificity and the stability of these characteristics over time. Two types of tests, *in vivo* and *in vitro*, can be used to verify the neutralizing power of the antivenin. Their principle is identical: it consists of measuring the lowering of the toxic effects of the venom, animal lethality or particular biological activity in the presence of increasing concentrations of antivenin.

For the *in vivo* tests the venom/antivenin mixtures are prepared in varying proportions and injected into the animal to evaluate the effect of the antivenin on the lethal effect of the venom or on a particular biological activity. One can work with a fixed quantity of venom (often defined as so many LD_{50}) and varying quantities of antivenin or the other way round. The venom/antivenin mixture is incubated at 37°C for a set time before being injected into the animal. The animals are placed under observation and in this way the quantity of venom is determined which leads to the death of 50% of the group, which corresponds to the quantity of venom neutralized by the antivenin. Often this quantity is expressed as number of LD_{50} per ml of antivenin. To be strict it would be preferable to express it as mg of dry venom per unit of volume with the LD_{50} referring to the trial animal and not necessarily to man or other animal species.

In medical practice the inoculation of the venom always occurs before the administration of the antivenin, sometimes with a long delay. In consequence sequential procedures have been proposed as well as other parameters like the fatal time

Antidotes and Immunotherapy

limit (FTL) or the last mortality time (LMT). These procedures are very suitable for the neurotoxic venoms. They permit the evaluation of the antigen/antibody affinity and the extraction capability of the latter, meaning its capacity to wrench the toxin off its receptor. However, they appear to be less well suitable for the venoms with toxic enzymatic activity, which cause a cascade of reactions running its own course once triggered, even after the enzyme has been neutralized. It is seen on one hand that the quantity of antivenin necessary to neutralize the lethal effect of the venom increases as the delayed administration of the antivenin approaches the fatal time limit and on the other hand that it becomes impossible to protect the individuals when this limit has been reached or passed.

However, when one wants to standardize the methods which determine the neutralizing power of an antivenin, the venom/antivenin pre-incubation is the most commonly used technique: the results are free of the influences of the toxicokinetics of the venom and of the pharmacokinetics of the serum and depend above all on the concentration of antibodies and their neutralizing capacity. It is possible to establish a neutralization curve of the antivenin with regard to increasing quantities of venom. This curve can be linear with each dose of venom being neutralized by a proportional quantity of the antivenin: from this one can conclude that only one toxic antigen is active or that the other toxic components, if they are there, are neutralized more efficiently (Fig. 30).

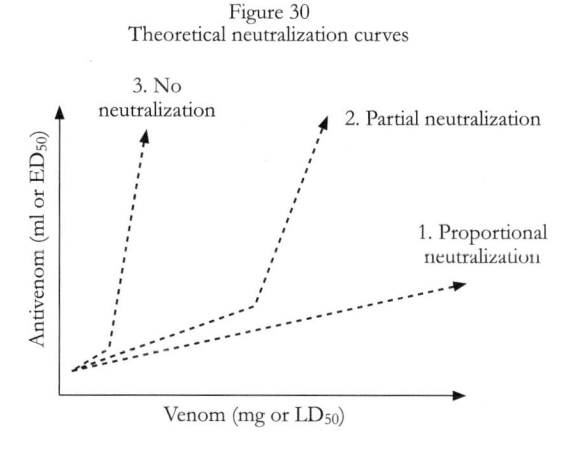

Figure 30
Theoretical neutralization curves

3. No neutralization

2. Partial neutralization

1. Proportional neutralization

Antivenom (ml or ED_{50})

Venom (mg or LD_{50})

Often the curve is bent, be it that one toxic component present in small quantity has for that reason induced few neutralizing antibodies or that a toxic component present in large quantity reveals itself as poorly immunogenic. This is the reason why one has to use at least two lethal doses of venom, when one uses the method of variable volumes of antivenin to determine its protecting titer. In this way one determines the volume of antivenin capable of neutralizing the lethal dose of the chosen venom: the seroprotection then is expressed as the dose of antivenin which reduces an elevated number of lethal doses, generally 3 to 10 LD_{50}, to one simple LD_{50} (ED_{50}).

Whatever procedure is used, the statistical calculations are done in the same way as for the determination of the LD_{50}.

The *in vitro* tests allow the evaluation of a particular activity of the venom and its neutralization after mixing the venom with the antivenin. Several such tests have been recommended for the determination of the neutralizing power of an antivenin. The neutralization of the haemolytic action of the venom tested on rabbit, sheep or human erythrocytes appears to correlate with the neutralization of the lethal action in mice. The fibrinolytic activity is evaluated on plates with sheep or cattle fibrin obtained by coagulating fresh plasma. The proteolytic activities are estimated from the action of the venom on various substrates, most often casein. The principal problem is to define a reproducible technique. However, in testing the capacity of an antivenin it becomes increasingly necessary to neutralize each toxin individually, particularly the neurotoxins and the myotoxins.

Quantitative ELISA tests can usefully add to the *in vivo* and *in vitro* tests. For this purpose micro-titration plates are lined with the venoms used for the preparation of the antivenin. The serum is incubated on the plate in varying concentrations and the presence of antigen-antibody complexes is revealed by an antibody directed against the immunoglobulins and conjugated with an enzyme that reacts on a substrate that releases a colorant (Fig. 31).

The intensity of the colour reaction as a function of the concentration of the antivenin and obtained after the addition of a fixed quantity of the substrate, can be compared to the DF_{50} or to the neutralized LD_{50}. The same can be done with the toxic fraction or the main toxin of the venom. This type of test can be developed if close correlations with the in vivo tests are found.

Figure 31
Mechanism of the ELISA test

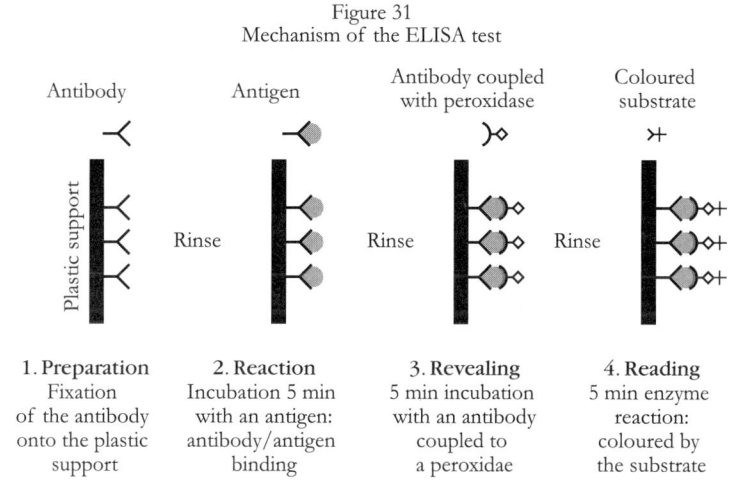

| Antibody | Antigen | Antibody coupled with peroxidase | Coloured substrate |

1. Preparation
Fixation of the antibody onto the plastic support

2. Reaction
Incubation 5 min with an antigen: antibody/antigen binding

3. Revealing
5 min incubation with an antibody coupled to a peroxidae

4. Reading
5 min enzyme reaction: coloured by the substrate

2.1.3 Stability of antivenins

The stability of antivenins is good, as long as certain precautions are taken. Liquid sera have to be stored in the refrigerator at +4°C, which is sometimes difficult in the tropics. However, they remain stable at ambient temperatures provided they are protected from light. The protective titer of an antivenin varies little after having been exposed to 37°C for one year. No bacterial development is seen. However, the antivenin looses progressively its clearness with time and temperature. The observed turbidity is caused by the coagulation of heterogenous proteins and formally prohibits the use of the antivenin, even though its protective power is unchanged: the injection of the precipitate seen in the antivenin carries the risk of causing a sudden activation of the complement thus provoking an anaphylactic shock. If kept under good conditions, liquid antivenins keep their properties intact for at least five years, lyophilized ones even longer: the laid down norms are well within the experimentally observed limits. The legal expiry time of liquid antivenins kept under refrigeration is three years and that of lyophilized preparations five years at ambient temperature.

The efficacy of ovine Fab does not appear to be altered after storage at 50°C for 60 days and only shows a slight decrease after 30 days at 70°C. However, shaking significantly reduces its activity.

2.2 Mode of action of the specific immunoglobulins

2.2.1 Antigen-antibody binding

The action of the antivenin is based on the contact of the antibody with the corresponding antigen. This can be modeled by a simple statistical problem of the two parts meeting each other in a dynamic biological system. The ideal would be that the antibody diffuses in the same environment and under similar conditions than the antigen. One also has to be certain that first neutralization occurs and then the elimination of the antigen-antibody complex. Neutralization of the antigen is a distinct property from precipitation. Neutralization causes a structural change that prevents the normal functioning of the native antigen. If the structural change affects the active site of the antigen, then its activity will be modified. One can distinguish two types of active antibodies, notably against the neurotoxins:

- The "protective" antibodies which recognize an epitope close to the toxic site of a free toxin, which prevents its binding on its receptor; however, once the toxin is bound to its receptor the epitope may no longer be accessible to the antibody. This function is confirmed by polyclonal antibodies and is exploited in immunotherapy.

- The "curative" antibodies can recognize an epitope distant from the active site and can bind to the toxin when it is already tied to its receptor, to wrench it loose from it; this property has been demonstrated with the help of monoclonal antibodies.

In practice this distinction is formal and probably there is a gradient of activity.

2.2.2 Cross-reactivity of antivenins

Cross reactions have been demonstrated by immunodiffusion and ELISA tests. They are seen frequently between related species. However, the existence of cross precipitation reactions does not necessarily mean that there is a cross protection. The antivenin against *Crotalus atrox* neutralizes the venoms of five rattlesnake species but not that of *C. durissus terrificus*, the only one characterized by the presence of a lethal neurotoxin, crotalin, which *C. atrox* does not have. A weak cross neutralization exists between the venoms of Asian and African cobras, except for

Antidotes and Immunotherapy

the venom of *Naja nigricollis*. The antivenins prepared from the venom of the Indian viper *Echis carinatus* are as effective against the venoms of the African vipers *E. ocellatus* and *E. leucogaster* as antivenoms prepared from the venoms of these two species. However, cross reactions are not necessarily reciprocal: The antivenin prepared from the venom of *Bothrops jararaca* reacts twice as strongly with the venom of *B. moojeni* than the antivenin of *B. moojeni* with the venom of *B. jararaca*. The cross reactions cannot really be foreseen and have to be verified individually and this makes it necessary to choose carefully the venoms to be used for the preparation of an antivenin.

2.2.3 Pharmacokinetics of antibodies

The composition of antivenins can differ by the degree of purification and above all by the nature of the antibodies which they contain: complete immunoglobulins (IgG) and immunoglobulin fractions (F(ab')$_2$ or Fab). These differences are very important with regard to the speed of diffusion and the distribution of the antibodies.

The apparent distribution volume of the antibodies in the form of IgG can be superimposed on the vascular compartment (cf. Fig. 26). In the tissues the IgG disperse much more slowly than the venom, as they have a strong affinity for the vascular compartment where they tend to remain. The IgG reach their maximum concentration in the superficial tissues in 6 hours and the deep tissues in 30 hours. The distribution volume of the F(ab')$_2$ is twice as high than that of the plasma volume, which suggests a better tissue distribution than that of the IgG, even though still mediocre. In addition they diffuse more rapidly than the IgG: in 1 hour in the superficial tissues and 6 hours in the deep tissues. Theoretically the Fab diffuse more rapidly in all the compartments than the other types of antibodies because of their small size. Their affinity for the deep tissues is five times stronger than that of the IgG and two times more than that of the F(ab')$_2$. However, their speed of diffusion differs little from that of the F(ab')$_2$.

In contrast to the venom which diffuses largely in the tissue compartment, the IgG and the F(ab')$_2$ diffuse only weakly outside the vascular department. However, the rapid decrease of venom plasma concentrations after the administration of the antivenin confirms that the serotherapy eliminates the venom not only from the general circulation but also from all the tissues (Fig. 32).

Figure 32
Neutralization of the venom by immunoglobulin fragments
(after Rivière and Bon, 1999)

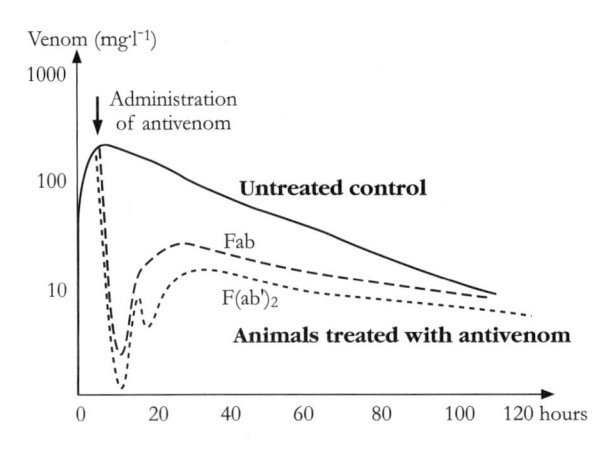

The antivenin causes the venom to be transferred from the tissue compartments to the vascular compartment following the law of mass distribution. This tendency to equilibrate the venom concentrations in the different compartments justifies the injection of a larger volume of antivenin than required for the neutralization of the venom, in order to accelerate the diffusion of the venom. This excess of antibodies helps on one hand to accelerate the transfer of venom to the vascular compartment and on the other hand to optimize the encounter between antigen and antibodies in the blood vessels.

Also the venom is eliminated from the body more quickly than the antivenin. In practice it appears that the life span of the F(ab')$_2$ is about 3 to 4 times longer than that of the toxins, quite sufficient to allow them to meet the toxins and to bind to them. The eventual failure of some immunotherapy treatments is therefore not linked to a too rapid elimination of the immunoglobulins but to a failure of the antigen/antibody contact because of an insufficient concentration of the antivenin or of a poor neutralization. This can be explained by a lack of specificity of the antibodies as well as by the persistence of functioning of the active site.

The half-life of the F(ab')$_2$ is approximately 50 hours. As the glomerular filter in the kidneys is impermeable to molecules with a molar mass above 50-60 kDa, antibodies in the form of IgG or F(ab')$_2$ are eliminated by the cells of the immune system. The fragments Fab are eliminated 10 to 20 times more quickly than the

IgG and have a five times better distribution in the vascular compartment. They are just on the threshold of renal filtration and are eliminated via the urine.

A significant superiority of the immunoglobulin fragments F(ab')$_2$ and Fab above native IgG has never been demonstrated. However, the use of whole IgG presents some theoretical disadvantages in comparison to the administration of immunoglobulin fragments. The quantity of protein injected is higher for the same level of efficacy and the fragment Fc of the IgG binds to the complement; these two factors bring about a considerable risk of side effects. The advantages of using the fragments F(ab')$_2$ and Fab over those of IgG have to be considered in the light of their respective disadvantages (Table XI).

Table XI
Comparison of the properties of immunoglobulin and its fragments

Properties	IgG	Fab2	Fab
method of harvest	precipitation	precipitation + pepsin	precipitation + papain
distribution	> 6 hours	3 hours	1 hour
elimination (half-life)	> 100 hours	60 hours	10 hours
tissue affinity	1	2	5
complement fixation	yes	No*	no
immunological affinity	1 to 2	1 to 2	1
excretion	immuno-competent cells	immuno-competent cells	renal

*activation of the complement alternative route

The quantity of proteins administered is similarly reduced in the two preparations. The F(ab')$_2$ does not bind on the complement but activates the alternative route of the complement, thus enhancing the release of C_3' with clinical consequences which are difficult to evaluate. On one hand the venom spontaneously activates the complement but on the other hand it releases interleukine 6, which provokes a synthesis of anti-stress proteins that protect against anaphylaxis. What ever is the case, the activation of the complement by the immunotherapy has never been demonstrated clinically. The fragment Fab is *a priori* better tolerated and its neutralizing power has been widely confirmed. In contrast, its excellent diffusion into the deep tissues appears to cause the immune complexes to remain at this level which retards their elimination. In addition the short half-life of free Fab reduces the length of their activity and imposes the need of repeated administrations that may cause side effects and be dangerous because of the higher doses of injected proteins. Finally the elimination of the antigen/antibody complexes becomes the main limiting factor. Taking place exclusively via the kidneys, the elimination of the complexes formed with the Fab can be blocked if they exceed the threshold size of glomerular filtration, which frequently appears to be the case with the proteins of snake venoms. The risk of kidney damage that is marked by a noticeable decrease of creatinine excretion, is therefore a real one with this type of immunoglobulin fragments. The not eliminated immune complexes remain in circulation in the cardio-vascular system and some of them can still have their toxic sites. The immune complexes formed with the $F(ab')_2$ are eliminated by the immune-competent tissues, liver, spleen and lymph nodes, which is quick and independent of their molecular weight and avoids nephro-toxicity.

2.3 Use of specific immunoglobulines – indications for the immunotherapy

The use of a monovalent antivenin, which in theory is more effective, depends on the formal identification of the snake that caused the envenomation. The normally used clinical criteria can be insufficient. If the identification of the snake is uncertain, a polyvalent serum has to be used against the main poisonous species known in the geographical region in which the envenomation took place. The fact that the polyvalent serum also has a cross-reactivity provides a further margin of safety.

2.3.1 Quantity of injected venom

This great unknown should determine the quantity of antibodies to be administered. The quantity of injected venom normally is estimated according to the intensity of the clinical symptoms. Their early appearance is generally a sign of severity. Because of the lack of precision of this method immunochemical tests have been proposed to quantify the venom in the blood and in the urine: passive haemagglutination, radio-immuno-titration, immuno-electrophoresis. These tests lack sensitivity with the exception of the radio-immuno assay (RIA), which is difficult to set to work (cf. Table XVI). The ELISA tests are able to detect the early presence of very small amounts of venom in the blood, the urine and even in organ samples. The relative ease of use and the short delay waiting for the results are important advantages in medical practice; kits for this purpose are presently being evaluated.

These tests can also be used to evaluate the efficacy of the antivenin. However, they are not entirely without faults. Non-specific cross reactions can occur or the sensitivity could be insufficient. In a general way there is an inverse relation between sensitivity and specificity: there is a higher risk of false reactions in the zones of high sensitivity.

In spite of these few drawbacks these tests have been used to verify whether there is a correlation between the quantity of circulating venom and the intensity of clinical signs. Scales of severity have been drawn up and compared with the quantity of venom in circulation and also sometimes with the quantity of venom excreted in the urine.

2.3.2 Indications for antivenin

Antivenin is the only specific medication directly able to neutralize the effect of the toxins present in the venoms: the principle of its use can therefore not be contested.

The indication of the antivenin will depend on the circumstances of the bite, the delay since the bite, the symptomatology, the medical environment and particularly the availability of an intensive care unit. In Europe antivenin is indicated in a child if the envenomation is certain and in the adult if it is severe and accompanied by haematological signs. In tropical countries the indication is wider, particularly in children and in pregnant women in the presence of clinical or biological signs pointing at an envenomation. Induced bites represent a particular case: the identity of the snake is certain, and the bite inflicted by an animal in a particular state of aggressiveness is potentially more severe. Also some patients are recidivists and therefore are exposed to reactions of sensitization to the venom or to the antivenin. The immunotherapy is not indicated if after an observation period of three hours no clinical or biological manifestation appears. The severity scales or scores established by different authors aim at defining and facilitating the indication for an immunotherapy (cf. Table XV).

For many the immunotherapy appears now as a global treatment of envenomations and no longer only as an antidote to the lethal effects of the venom; the antivenin prevents the development of local and haematological complications like the thromboses seen in cases of envenomations by *Bothrops lanceolatus*. It reduces the time of hospitalization of the patient, in itself already an important indication.

2.4 Optimization of the therapeutic protocol and administration of the antivenin

The rules for the administration of the antivenin remain controversial regarding the delay beyond which the immunotherapy looses its efficacy, the route of administration and the posology.

2.4.1 Delay of administration

It is agreed that immunotherapy, once the indication is given, is the more effective the earlier it is administrated. However, a long delay between the bite and the start of treatment does not lead to the exclusion of the immunotherapy. The same dose of antivenin given at different times after the injection of the venom leads to an identical level of neutralization (Fig. 33).

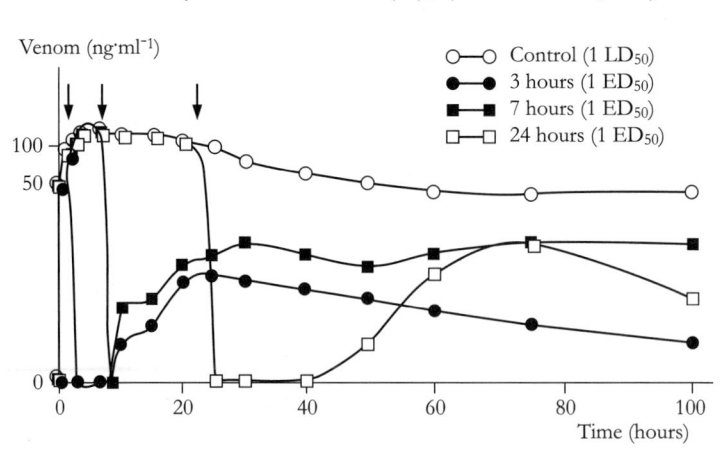

Figure 33
Effects of delayed administration of F(ab')$_2$s (after Rivière *et al.*, 1997)

Antidotes and Immunotherapy

As long as the damage that has occurred before the treatment is not definitive, the efficacy of the serum therapy, even if late, can be significant. In addition, in some cases the anti-toxin antibodies are able to destabilize the cellular toxin-receptor bond. Also the clinical signs appear slowly, like the hemorrhages after bites by *Echis* spp., due to a defibrinating action of the venom, which are severe but appear slowly. Recovery without sequels has been reported in patients bitten by viperids and treated after delays. It is not possible to determine a time limit after which an immunotherapy no longer has an effect on the envenomation, but the posology has to take into account the delay before the first administration of the immunotherapy and must be adjusted as a function of the clinical state.

2.4.2 Route of administration

In view of the potentially more rapid diffusion of the toxins with their generally lower molar mass than that of the antibodies, the intravenous route is presently recommended by most authors. This has also been proven experimentally. The speed of diffusion and of the bio-availability are significantly greater than by other routes, particularly the intra-muscular one, which is up to now recommended by many producers of antivenins. The dose necessary to neutralize 1 DL_{100} is four times higher when given intramuscularly then when given intravenously. The intra-muscular injection is not very effective and does not avoid the adverse reactions. The subcutaneous injection around the site of the bite should be prohibited: it is painful, ineffective and can induce local complications without avoiding the general side affects.

Most often the antivenin is given by slow perfusion, diluted in an isotonic solution. The slow, direct injection allows one to reduce by more than half the injected quantity for an equivalent efficacy and a more rapid clinical improvement. Also, the perfusion allows one to adjust the administration of the antivenin and to facilitate the control of eventual adverse reactions.

2.4.3 Posology

It is based on the diagnoses of the snake responsible for the envenomation taking into account the mean glandular capacity, on the delay before the start of the immunotherapy, on the clinical state, on the titer of the antivenin and on the medical environment. Age, sex and body mass of the patient are not important criteria.

In the absence of an up-to-date facility for the evaluation of the amount of toxin in the circulation, one tries to have an excess of antibodies for the elimination of all free toxin from the circulation in the body. The saturation of the vascular compartment with $F(ab')_2$ leads to a complete and durable neutralization (Fig. 34).

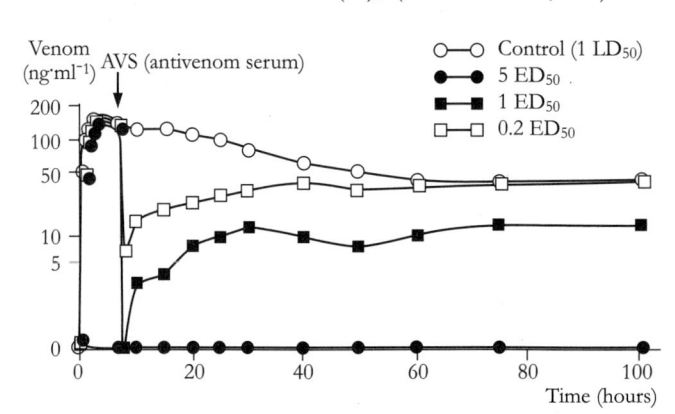

Figure 34
Effects of different doses of $F(ab')_2$s (after Rivière *et al.*, 1997)

In contrast, insufficient doses show rapidly their ineffectiveness, even if they transiently neutralize the venom, by the re-appearance of venom from the deep tissues. However, beyond a certain quantity of antivenin the therapeutic benefit becomes negligible. The scales of clinical severity can serve not only for deciding on the indication for the immunotherapy, but also for better adapting its posology. Doses of 100 to 150 ml given in one day are recommended in the case of tropical snakes. Presently much lower dosage rates are under consideration.

2.5 Adverse reactions caused by antivenin

The side effects observed during the immunotherapy are due to the administration of foreign proteins, the previous sensitization of the patient to horse serum or the presence of immune complexes that are difficult to eliminate by the body (Table XII).

Antidotes and Immunotherapy 155

Table XII
Classification of the side effects caused by immune therapy

	antibody	antigens	Cells	mechanisms
HS I	IgE	non-specific	mast cells	release of histamine
HS II	IgG/IgM	cellular	K cells	activation of the C' (cytotoxicity)
HS III	IgG/IgM	soluble	polynuclear	activation du C' saturation of the immune complexes
HS IV	-	variable	T lymphocytes	release of lymphokines

HS = hypersensitivity

The first of these effects are the non-specific type I hypersensitivity reactions that are in proportion to the amount of injected proteins. They appear within minutes of the administration of the antivenin. The sensitization to horse proteins corresponds to a type III or VI hypersensitivity response with immediate (less than 12 hours) or delayed reactions (1 to 3 weeks). Generally these reaction are benign, particularly the type I hypersensitivity reactions. However, sometimes they can be sudden and severe as in anaphylactic and anaphylactoid shocks. Finally the type II hypersensitivity reactions are specific and caused by the venom itself. They need a previous sensitization, a previous contact, and thus occur in persons who had already been bitten or those who handle venom. These are generally violent anaphylactic shocks, potentially fatal in the absence of appropriate treatment.

2.5.1 Early effects

They occur in sensitized patients who previously had received an antivenin or antitoxin (e.g. anti-tetanus serum) immunotherapy, but also in patients who had no previous immunotherapy. In the first case one speaks of anaphylactic reaction and in the second one of anaphylactoid reaction. The presence in the antivenin of a strong proportion of Fc fragments, which have no antibody activity but activate the complement, can induce an anaphylactoid shock, but this can also be caused by the venom itself. The frequency with which these accidents occur is estimated variably. Some poorly purified antivenins can lead to up to 30% of side effects or even more. However, the sudden, life threatening anaphylactic shock appears to be rare, particularly with purified immunoglobulin fragments. In addition to differences in the nature of the commercial antivenins, the variation of incidence and intensity of side effects can also be explained by the perception and interpretation of the symptoms by the patient or the physician.

2.5.2 Delayed side effects

They are less well known than those discussed above. The heterologous antibodies of the antivenin need approximately three weeks for their elimination from the body, which during this time produces its own antibodies against the not yet eliminated serum. In a certain number of cases the precipitating complexes are formed quickly and from this derives the name "9th day disease" sometimes given to this late reaction which is also known as serum disease or immune complex disease. The precipitating complexes are deposited on the walls of the small blood vessels and provoke a series of usually moderate symptoms, mainly fever, urticaria, adenopathies, arthalgias and nephropathy with proteinuria. The severe forms, a glomerulopathy with neuropathy, are the exception. Generally the course is benign with a spontaneous recovery within two to four days. Generally a treatment is not necessary. If the symptoms are severe, corticoids or antihistamines can be given. However, this type of accident, like the preceding one, can constitute a contraindication for a later immunotherapy.

Skin and conjunctiva tests have been proposed to try to recognize patients likely to suffer from intolerance reactions to antivenin. Generally their predictive value is regarded as unsatisfactory. Also a positive test does not preclude an immunotherapy if it is required by the clinical state of the patient. Consequently these tests tend to be abandoned. The incidence of side effects, whatever their nature or their intensity, is estimated variously but would be below 5% of patients treated with the new generation of antivenins composed of highly purified immunoglobulin fragments. With these products the incidence of serious and severe side effects now lies far below 1%.

In spite of the reduction of the risk of side effects precautions have to be taken systematically: availability of appropriate drugs (adrenalin, corticoids) and posttherapeutic observation.

2.6 The future of immunotherapy

Several solutions can be envisaged to improve the immunotherapy with the aim to obtain a greater efficacy and a better safety of use.

2.6.1 Improvements of the immunization of the animal

Antigen quality

The quality of the venoms used for the immunization must be monitored rigorously. A logical choice of venoms serving for the preparation of antivenins should be based on the combination of a large number of fractions and a low toxicity. The use of toxic fractions isolated beforehand can increase the protective titer of the antivenin and reduce the quantity to be injected while at the same time avoiding the formation of antibodies against non-toxic proteins.

Purified toxins can be used to enrich a venom which is poor in toxins and even to substitute it: it has been shown that the antibodies against crotoxin, the principal toxin of the venom of *Crotalus durissus terrificus*, can neutralize the effects of the complete venom.

Through cloning the toxins of elapids or viperids one can envisage a production with the same aims. The production can be perfected by trying to select recombinant variants that are less toxic and more immunogenic.

Finally, attempts to produce venom in vitro by culturing cells of the venom glands are still in the experimental stage.

2.6.1.2 Immunization procedures

In addition to the production of anavenoms and anatoxins by various physicochemical processes or by molecular biology techniques, other processes have been proposed, in particular the use of liposomes. The venom is incorporated into an emulsion of sphingomyeline and cholesterol stabilized in membrane form. The administration of such preparations by oral or parenteral route is followed by a rapid rise in specific protective antibodies. The subcutaneous route appears to be the most effective one, as the antigens remain in contact with the tissue macrophages, which are the most efficient cells to present the antigens. The slow release of the antigen from the liposomes and its diffusion via the lymphatics allow the continuous stimulation of the immune system. The presence of bacterial liposaccharides in the liposomes can improve the immunological response. However, because of its toxicity it cannot be used in man. Other adjuvants have been tested; their efficacy varies with the animal species that is used and can be negligible, particularly in the horse.

2.6.2 Improvement of antibody quality

2.6.2.1 Purification of antibodies by immuno-affinity

This technique would allow the retention of the specific antibodies only. The purification of sera by immuno-affinity can be carried out on gels that are chemically coupled to the antigens of the venom. Proteins that do not react with the gel are eliminated from the preparation, while the specific antibodies are retained. Thereafter these are recuperated by a change of pH, which destabilizes the antigen-antibody bond. The antibodies harvested in this way are specific for the venom used and have a higher specific neutralizing capacity. The doses to be administered can therefore be smaller, which can help to reduce the incidence of secondary reactions. However, the cost of production is higher. There is also the risk of inadvertent contamination by the venom liberated from the coupling gel.

2.6.2.2 Isolation of specific immunoglobulins

The isolation of the IgG_T, which are the principal IgG of the horse involved in neutralizing the venom, would reduce the quantity of injected heterologous proteins. However, the IgG_T are strongly glycosylated and can cause allergies, which leaves part of the problem unresolved.

2.6.2.3 Fragmentation of antibodies

The controlled enzymatic digestion of the immunoglobulins allows to obtain different types of active fragments. The respective advantages and disadvantages of the $F(ab')_2$ and Fab have already been discussed (cf. Table XI).

It is also possible to prepare the Fv fragments that are dimers made up by the combination of the first domain of the heavy and the light chain of the IgG. First obtained by chemical cleavage they have lately been synthesized by molecular biology techniques. Their pharmacokinetic characteristics have not yet been established, but they offer interesting possibilities. Molecular biology also allows the construction of the single chain or scFv (single chain Fragment variable), variable fragments of antibodies, but their experimental use in immunotherapy has not yet been tried. It is not yet certain whether they have real therapeutic advantages, particularly if one considers the production constraints. In contrast they can be used in the identification and titration of circulating venom in an envenomation.

2.6.2.4 Production of monoclonal antibodies

At first view the use of monoclonal antibodies offers an alternative solution to the use of polyclonal antibodies presently in use. They are obtained after fusion of spleen B lymphocytes of hyper-immunized mice with murine myeloma cells. The hybrid or hybridoma cells have the potential of unlimited multiplication and secrete antibodies of determined specificity. Monoclonal antibodies derived from a single cell clone can combine only with one single epitope. In contrast the polyclonal antibodies produced by multiple cellular clones can combine with most of the antigenic determinants of the antigen. A certain number of neutralizing monoclonal antibodies has been described and the neutralization of the lethal power of the toxins by these antibodies has been studied. A monoclonal antibody against the post-synaptic neurotoxin of *Naja nigricollis* causes the dissociation of the complex formed by the toxin with the post-synaptic acetylcholine receptor. They then combine in a ternary complex receptor-toxin-antibody that has no toxicity. A monoclonal antibody which *in vitro* inhibits the enzymatic activity of crotoxin, the phospholipase A_2 neurotoxin of *Crotalus durissus terrificus* acting on the pre-synaptic level of the neuro-muscular junction, also neutralizes *in vivo* the lethal power of this toxin. The murine antibodies can cause an immune reaction in man. One has tried to reduce this reaction by "humanization", by transplanting the hypervariable regions of the murine antibody, which carry the neutralizing function, onto constant regions of a human antibody. Such chimeric molecules, produced by molecular biology techniques, have already been realized against lymphocytic receptors or in antiviral therapy. They have better pharmacokinetic parameters than the murine monoclonal antibodies and do not appear to provoke an immune response. These processes can be applied to the Fab or the Fv. These techniques are not only costly but the use of monoclonal antibodies for the treatment of human patients is contra-indicated as long as they are produced by tumour cells.

2.6.2.5 Change of animal species

Other species than horses are suitable for the production of antivenins for therapeutic use: in this way hypersensitivity reactions due to a previous injection of equine serum can be avoided. They have amongst other advantages the one that they do not produce sensitizing IgG_T. Some commercial antivenins have already been prepared from sheep (Annex 2). However, the non-conventional transmissible agents (NCTA) throw a shadow over the present way of antivenin preparation; as long as many unknowns remain with regard to the mode of transmission of these conditions and their early detection is not solved in a simple fashion, one hesitates to recommend the wider use of antivenin derived from hyper-immunized sheep. In the present state of knowledge it will be better also to avoid bovines, which in any case are rarely used for therapeutic sera. There are some goat antivenins, anti-*Bungarus multicinctus*, anti-*Calloselasma rhodostoma*, and anti-*Vipera latasti*, or prepared from rabbits (anti-*Naja naja*).

In the particular context of the NCTA there is also the possibility of obtaining neutralizing antibodies from birds. They can be harvested easily from the eggs. The avian antibodies have a high protective power and do not react with human complement, which reduces the frequency of side effects. However, the IgG of the fowl are strong allergens and the general yield could be much lower than from the horse. One could also wonder about the possibility to study the tolerance and the yield from larger birds, geese, turkeys, ostriches, as their production is being developed and their eggs are underutilized.

Recently the attention has been drawn to the immunoglobulins of the camelids. Some of the immunoglobulins of the camelids lack a light chain, which on one hand reduces their molecular weight and on the other hand enhances the recognition of the epitope.

2.6.3 Improvement of treatment protocols

The treatment of envenomations is not standardized and the practices are largely empiric, not very rational, ineffective and even dangerous (subcutaneous injection around the bite, under-dosage in children, posology independent of the severity of the envenomation or clinical problems).

2.6.3.1 Adaptation of the posology

The immunological titration of the venom and of the antivenin in circulation is already possible with the help of ELISA tests. The limits of these tests have been mentioned. Yet they offer the possibility to better adjust the quantity of antivenin needing to be administered. In the absence of this possibility one relies on the symptoms and their development to evaluate the severity of the envenomation and to deduce the quantity of immunoglobulins needing to be administered and the need of their repetition.

2.6.3.2 Potentiating the antivenin

This allows to reduce the quantity of antivenin to be injected and thereby the side effects and the cost of the treatment. An effort is made to find adjuvants which re-enforce the neutralizing power of the antivenins and which could have an antagonistic affect on the venoms. The effect of certain drugs or plant substances on envenomations has been known for a long time. In some cases the theoretical benefit of such a substance has been confirmed in clinical experiments.

3. Medical Treatment

3.1 Treatment of the neurological problems

The neurotoxins that bind to the cholinergic receptor of the postsynaptic membrane are antagonized by neostigmine. Neostigmine acts by inhibiting cholinesterase, the enzyme that destroys acetylcholine after its action on the receptor of the postsynaptic membrane. In a cobra envenomation the administration of neostigmine keeps the acetylcholine on its receptor authorizing the passage of the nerve signal and compensating the blockage of the latter caused by the binding of the neurotoxin on some of the receptors. In other terms, neostigmine enhances a competition between the physiological mediator of the nerve stimulus acetylcholin and the neurotoxin that hinders the conductivity of the nerves. The limits of a treatment with neostigmine will therefore depend on the number of receptors left unoccupied by the neurotoxin at the time of the treatment. It is thus important that the treatment be given early unless it is given in conjunction with an immunotherapy freeing the cholinergic receptors that are immediately occupied by the neostigmine. In a single dose neostigmine triples the delay before the onset of the lethal effects of the venom, while the same dose in fractionated amounts increases the delay by five times.

The 4-aminopyridin and its derivatives block the potassium channels and enhance the penetration of calcium into the pre-synaptic cells and stimulate the release of acetylcholin. These substances increase the excitability of the nerve fiber and prolong its action potential. These potassium channel blockers are more particularly active on the diaphragmatic synapses. One therefore has a β-neurotoxin antagonist that is probably less active against an intoxication by α- neurotoxins.

The facilitating toxins or dendrotoxins of the venoms of *Dendroaspis* are antagonized by atropine. These toxins act on the postsynaptic membrane and enhance the release of acetylcholine into the synaptic gap. These repeated and intensive discharges of acetylcholine cause an excitation of the parasympathetic receptors in the heart muscle fibers, the smooth muscles of the digestive tract and of the lungs and of the central nervous system. Atropin binding to the same receptors prevents the acetylcholine from binding. In contrast to neostigmine, the effect of atropine does not depend on an early treatment. In effect, the action of acetylchline is short because of the action of acetylcholinesterase. The effect of atropine on the venoms of elapids, which have no facilitating toxins, aggravates the respiratory problems while potentiating the curarizing neurotoxins: There is thus a real danger if they are not used with discretion.

3.2 Treatment of local symptoms

In the case of the venoms of viperids the anti-inflammatory and analgesic effect of many drugs will be of particular use. However, substances acting on blood coagulation have to be avoided as they can aggravate the haemorrhagic syndrome that also occurs in viperine envenomations.

The salycilate anti inflammatories will have to be avoided *a priori* because of their action on the blood platelets and on haemostasis. The corticoids have a powerful anti-inflammatory action and cause quite an enzyme inhibition, which can be beneficial in some viperine envenomations. One has to take into account the contra-indications of the corticotherpay, particularly the haemorrhagic complications and the immune suppression, which can accentuate the risks of a local infection.

3.3 Treatment of the haemorrhagic syndrome

The treatment of the haemorrhagic syndrome is much more difficult.

The sulfuric polyester of pentosane (Hémoclar®) has shown experimentally and clinically a certain efficacy although the reason of this has not been elucidated.

Heparin and its derivatives are often recommended but appear to be of use only in some particular instances. Generally, as we have seen already, heparin has no inhibitory effect on the thrombin-like enzymes of the venoms of viperids. However, one cannot exclude the fact that experimental administration of heparin reduced the toxicity of certain venoms.

Heparin also reduces the myotoxic effects of the venom of *Bothrops jararaca* and increases the efficacy of the antivenin. Drugs that inhibit the enzymes of the fibrinolytic system, tranaxemic acid (Exacyl®) or ϵ-amino-caproic acid (Hemocaprol®) can be prescribed when a primary or secondary fibrinolytic activity is suspected. It is probably also that property which explains the efficacy of the corticoids. Certain inhibitors are being evaluated either for local or if their tolerance permits for generalized application. Experimentally the efficacy of antithrombin III in synergy with the antivenin has been demonstrated against certain viperid envenomations. Antithrombin III would act by inhibiting certain coagulating substances of the venom even when the antivenin is found to be insufficient. Experimentally EDTA has an indisputably inhibiting action on many venom enzymes, notably the metallo-proteases. Human clinical trials confirm this observation. Para-bromophenacyl bromide (Batimastat®) is an inhibitor of the metallo-proteases and its efficacy in local lesions due to the venom of *Bothrops asper* has been confirmed experimentally.

The thrombolytic drugs like streptokinase or urokinase could be of interest in certain envenomations, provided they are administered very early, at the moment when the action of the venom is too coagulant and can provoke visceral embolisms. The efficacy of these drugs still is controversial.

3.4 Supportive treatments

Finally, the action of some drugs remains unexplained. The antihistamins reduce the toxicity of the venoms and potentiate the effect of the immunotherapy. It is possible that the mode of action should be researched at least in part at the level of the inflammatory response that would be regulated and limited as far as its calamitous effects are concerned. However, the fact that the toxicity of elapid and viperid venoms is reduced in similar proportions leads to the thought that either a direct effect on the venom or an immune stimulating effect could be the cause of that.

The lethal effect of several snake venoms can be reduced by certain common drugs like chloroquin and chlorpromazine with regard to the venoms of *Bungarus caeruleus* and *B. multicinctus* or dexamethasone against the venom of *Oxyuranus scutellatus*. Other drugs like diltiazem, nicergoline, verapamil, also give a protection against the lethal effect of these venoms.

Diuretics accelerate the elimination of the toxins which are not bound by antibodies.

Experimental results confirm the role of certain plants used in traditional therapeutics and their different properties can become useful for symptomatic treatment (cf. Table IX). The conditions of their use still have to be defined to guarantee their tolerance and their efficacy and the necessary human clinical trials will have to be carried out.

4. Vaccination

Vaccination has been envisaged as a replacement of the immunotherapy. Technically, effective anatoxins are available which can be injected or given in the form of active liposomes by the oral route. In Japan the clinical analysis of *Protobothrops* bites has let to the conclusion that the victims of severe envenomations due to the introduction of massive doses of venom into their body, could benefit from an immunity, even a weak one, against the venom; it would delay the appearance of symptoms and allow a more efficient treatment. The haemorrhacic factors have been purified, inactivated by formalin and injected with adjuvants into volunteers. A sufficient antibody titer was reached after the third injection. However, morbidity and mortality remained unchanged and the trial was abandoned as a failure. Recently trials were carried out in Myanmar against the envenomation by *Daboia russelii*. The immunization of the volunteers was achieved with three doses with acceptable side effects. The immunological titer appears to protect against one lethal dose of the venom. The duration of the protection and the efficacy of the vaccine have not been evaluated for ethical reasons.

The main difficulty encountered with vaccination is the delay between the penetration of the antigen and the immune response. In the case of a microbial agent the pathogenic effect is delayed by several hours at least and generally by several days, which leaves time for the antibodies to be mobilized against the intruder. In the case of an envenomation the toxic action is immediate and leaves little time for the immune response and the production of a sufficient quantity of antibodies.

The benefit of the vaccination of part of a population must be measured against the risks of side effects for all the vaccinated subjects. In addition many problems remain linked to a large-scale vaccination of a population, notably in developing countries.

PART III
Envenomations

CHAPTER 5

The Epidemiology of Envenomations

The frequency of bites and their severity depend at the same time on the human activities and those of the snakes. The circumstances of the encounter between man and snake are not by chance and several factors contribute to the risk of snakebites. In addition to occasional or accidental bites, illegitimate bites linked to the voluntary manipulation of snakes have been observed for several decades and particularly in the industrialized countries.

1. The Ecology of Snakes

The activities of snakes are difficult to quantify. For the epidemiologist two types of information are important and often sufficient, the density of a population and its composition. The population density of snakes varies in space and time and explains to a large part the incidence of bites. The taxonomic composition, which is the proportion of the various species encountered in a precise area, has an effect on the severity of the bites.

Different methods are used to collect snakes and to study their demography, ecology, physiology and ethology. The techniques of observation depend on the objectives and have variable constraints. Most often they are used in combination.

- *Active capture* can be by opportunity or targeted and consists of collecting the snakes dead or alive as they are found. This is carried out along a natural transect (forest edge, river bank, cliff) or an artificial one (fence, road, railway line). The result can be measured as unit of surface, distance or time. The collected information is of demographic, ethological or physiological nature. A chosen method may introduce a large but often unforeseeable bias. The choice of place and time is selective and has a strong influence on the interpretation of the results. However, it can also be perfectly representative of a given epidemiological situation (notably occupational activity) and express a determined potential risk.

The exhaustive capture in a delimited area probably is the only suitable method for the determination of the real population density. In practice the area is defined and closed in by a fine mesh net and all the snakes within are caught.

- *Trapping* in its different forms is a passive form of collection derived from the preceding one. It consists of placing the devices designed to attract and hold the snakes. There are two types of traps: the shelter that simulates more or less the natural resting place of the snake without retaining the snake by force, and traps that prevent the animals from escaping. Trapping can have the same methodological limitations as active capture. Particularly the type of trap, the material used, the spot where it is placed and the frequency of visits all affect the results and their interpretation. The selection of animals to be caught cannot be controlled any better than is the case with active collection. In addition the devices are often costly and attract attention and many interferences, intentional or otherwise, sometimes malicious, which hinder the observations.

- *Marking* is a more rigorous and informative method. It can be done with the help of paint, rings, tattoos cuts on the scales or radio-active isotopes. These methods are simple but some of them are temporary or traumatizing; all can have an effect on the behavior of the animal or on facilitating its capture by predators. A recent technique consists of placing a small transponder of only a few mm^3 under the skin of the animal with the help of a syringe. In all these cases it is necessary to handle the animal for marking it and later for its identification, which may cause an observational bias and can be dangerous. Also the animal must be found again, which is another limiting factor. Since a few years there are small radio transmitters that send a continuous signal on a constant frequency and the snake can be forced to swallow such a transmitter or it can be implanted surgically (telemetry or radio tracking). Although this procedure is more costly, it renders any further contact with the snake unnecessary and does not oblige the investigator to make visual contact with the snake as the transmitted radio signal is sufficient to determine the exact whereabouts of the snake every time that this is required. In this way the observer does not interfere with the movements of the snake. These techniques require a large investment of material and/or time but they furnish detailed information. They can serve to give precise information on the movements of the snakes (distance, duration, causes, circumstances) and certain physiological (growth, longevity, sexual cycle), ethological (prey, predators, day-night cycle) and demographic parameters (abundance, fecundity, mortality).

The marking and recapture technique, which consists of marking simultaneously a large number of animals that are then released to be recaptured later, is one of the most often used procedures for demographic studies.

1.1 Demography

The population density can be expressed in three different ways.

- *The absolute* (or real) *density* is the exact number of snakes per unit of surface area, of volume, or in some particular cases of linear measure (length of rivers or of hedges). This is an ecological indicator that is difficult and costly to obtain. It is of great importance for the demographic study of a population. However, it does not address the complexity of the circadian cycles of the snakes, which are active at variable times during the day and the night. For the epidemiologist this information is often more important than the actual density of a population.

- *The apparent density* answers this need. It is less precise than the absolute density as it depends strongly on environmental conditions, on the behavior of the snakes and on the research methodology. It corresponds to the number of snakes collected during a defined activity over a limited surface. This number is placed in relation to 100,000 potential catchers and can then be compared with the number of observed bites.

- *The abundance* corresponds to the number of individuals in a population. The abundance figures produced by marking and recapture are probably the most reproducible values and come closest to the real density while still leaving the possibility of an epidemiological interpretation. The principle is based on the frequency of capture of marked individuals in relation of the total result of the collection. The approximate total size of the population can then be figured out with the help of a simple formula.

Principally the absolute and apparent densities are proportional in the same location provided that the relation between the controlled parameters (number of catchers/number of exposed persons, time of capture/time of exposure, surface areas) is taken into account.

The differences in density allow the comparison of the abundance data between sites or more often at different periods. In this way it is possible to evaluate the effect of a natural or artificial event on a population or more generally the surveillance of the demographics of a population.

1.2 Abundance and apparent density

The demography is linked intimately to fertility and mortality. The numbers of progeny and the longevity differ very much between species. Vipers and cobras live on average between 5 and 15 years, crotalids 25 to 30 years and boas and pythons 50 years and more. The colubrids can live for 5 to 30 years.

The apparent density varies with space as land use of the area (plantations, division into lots) creates different environments with a direct influence on the behavior of the snakes. Depending on the species and the ecological changes taking place the snakes are either attracted or repelled. The apparent density also fluctuates with time under the influence of seasonal and physiological factors that affect their behavior.

Accumulations of snakes can have natural (hibernation, mating), accidental (inundations) or circumstantial causes (land use). Snakes do not have a developed social behavior. However, the concentration of individuals can become necessary by environmental constraints or by localized but dispersed favorable conditions. In temperate climates the snakes have to adapt to the climatic variations and to the rigors of winter. Considerable numbers of snakes can come together in a communal burrow. These shelters can be natural cavities that are unoccupied or burrows shared with other animals. Different species, even prey animals together with their predators can thus find a place in the same refuge. It is possible that these accumulations help the individuals to maintain their thermal equilibrium. At the end of winter the reproductive *Vipera aspis* attracts several males that intertwine and form a ball or knot, and this might increase the chance of her being fertilized. *Thamnophis sirtalis* is known for the accumulation of several thousand, up to 8,000, individuals in a winter shelter, which appears to enhance conditions for mating in spring. The nuptial parade can assemble several tens of males around one single female. As the same thermal conditions are required for gestation, there is also a tendency for the females to gather in a favorable place. It appears that this is rather because of the opportunity than of a real need to be together. For the same reason the females may also lay their eggs in the same place. Eggs of different clutches and even of different species can be found together in such a place.

The variations in population density can have natural, accidental or circumstantial causes. A lowered rainfall is accompanied by a strong decline in snake populations, because of the decrease of prey animals as well as because of unfavorable environmental conditions, notably the dryness of the soil. Bush fires used in developing countries for the practice of slash and burn cultivation cause a significant reduction of snake densities in comparison to uncultivated areas (Fig. 35).

In a general fashion the population density of snakes and their population diversity are linked as much to the abundance of prey as to conditions offered by the physical environment: natural hiding places, surface water, etc. and these latter have an influence themselves on the population density of prey species. Consequently one finds a progressive and significant

Figure 35
Comparison of absolute snake densities in burned and unburned savannah (after Barrault, 1971)

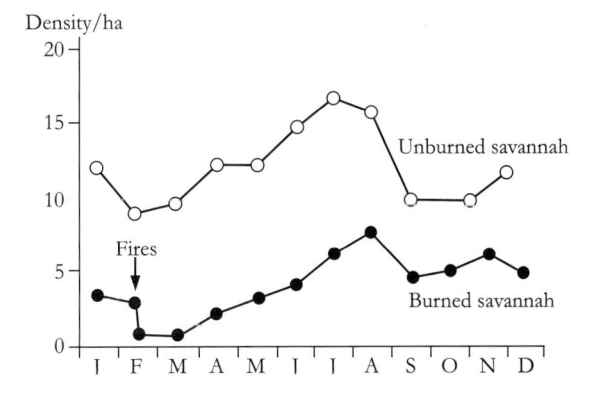

decrease of the snake population density and of the number of species from the forest to the Sahel. The destruction of snakes by man is far from negligible as cause of their decline. This destruction can be intentional (hunting poisonous snakes, but particularly for food and for their skins) or involuntary (road kill). Human population increase has a negative effect on most snake species even though it may be beneficial for some of them.

The population density varies with species and with geographical regions. The population density of the Mediterranean colubrid *Natrix maura* is about 10 individuals per hectare in the Massif Central and 30 in the plains of Languedoc. That of *Malpolon monspessulanus*, the Montpellier snake, is 10 individuals per hectare. The population density of the Far Eastern colubrid *Rhabdophis tigrinus* can be more than 42 individuals per hectare. On the island of Guam (Micronesia) densities of 80 to 120 individuals per hectare have been reported. Some species of colubrids even reach densities of several hundred individuals per hectare. Most often this happens at certain points, which can be explained by particular local conditions.

In venomous snakes the numbers are often smaller: 5 to 10 *Agkistrodon contortrix* per hectare in North America or *Protobothrops flavoviridis* on certain Japanese islands. On the Amami islands the density of *P. flavoviridis* is between 1 and 8.2 individuals per hectare, it is 11 specimens per hectare on the island Minnajima and 32 per hectare on Okinawa. In some cultivated areas it can reach some 20 individuals per hectare while in residential zones it is on average 2.5. The distribution of the population is, in fact, very irregular and next to areas entirely free of snakes one can find localities with several dozen individuals on a few hundred m^2, which brings the gross density to more than 600 crotalids per hectare. In northern Europe (Finland and the UK) densities of 5 to 15 vipers per hectare are observed. In southern Europe, notably in France, larger densities are sometimes mentioned. In tropical countries, South East Asia and sub-Saharan Africa, zones with several hundred snakes per hectare have been described. Such is the case of *Echis ocellatus* in the Sudanese savannah where the density per hectare can reach several dozen specimens.

1.3 The territory

Few facts are known on the territories of snakes, but the sum of observations tends to prove that the space in which a snake moves is limited and constant throughout its life. Probably this space is in proportion to the snake's biomass, namely its size and its energy requirements. However, some circumstances can push the snake to leave its normal territory. Environmental changes, whether natural (flooding, fires) or intentional (construction, land use), are reasons for leaving the territory. Climatic constraints (rainy season in the tropics or winter in the temperate zones) and human activities, also often seasonal, equally influence movements of snake populations. However, communal burrows can be far from the individual territory and oblige the snake to move quite far in spring to regain its territory: 350 m for *Thamnophis sirtalis* males, but up to 18 km for females, 1.5 km for *Vipera berus, Agkistrodon contortrix* and *Crotalus horridus*, 2 to 4 km for *Crotalus atrox* and *C. viridis*. However, a migration in the strict sense has never been observed except in certain marine snakes.

The configuration of the territory is adapted to the biotope: linear in hedgerows or mountains, while in forests, scrublands, heaths, fields and fallows it is polygonal and larger. To measure the surface of a territory is quite difficult: Depending on the mathematical method, an exaggeration can double or triple the really utilized surface. In effect, snakes generally move along the same paths, following their own spoor.

The territory changes little during the life of the snake. The translocation of snakes causes a five times higher mortality than normally expected. Displaced snakes show an abnormal behavior: They move more often and over longer distances than those that stayed in their original territory. The daily distances are increased threefold and the territory tenfold, which also confirms that the type of calculation artificially increases the surface of the territory. In subsequent years these movements tend to become normal again. Even in temperate countries after hibernation the snake finds again the place it had occupied the preceding year. However, some species can change their territory without being forced to do so by major ecological constraints. Such is the case with *Vipera aspis,* which in this way shows it great ability to adapt to very different ecological and climatic environments. These changes are relative, though; the newly occupied territory always includes a part of the old one.

The territories of *Vipera aspis* and *V. berus* vary between 300 m^2 and 2 ha depending on the landscape. On average it is 5,000 and 6,000 m^2. The daily distance covered by a viper is about 10 m on average. A male moves about three times further than a female. The females move infrequently, over short distances and only for short times with long periods of inactivity in between, while gravid females move even less frequently and over still shorter distances (Fig. 36).

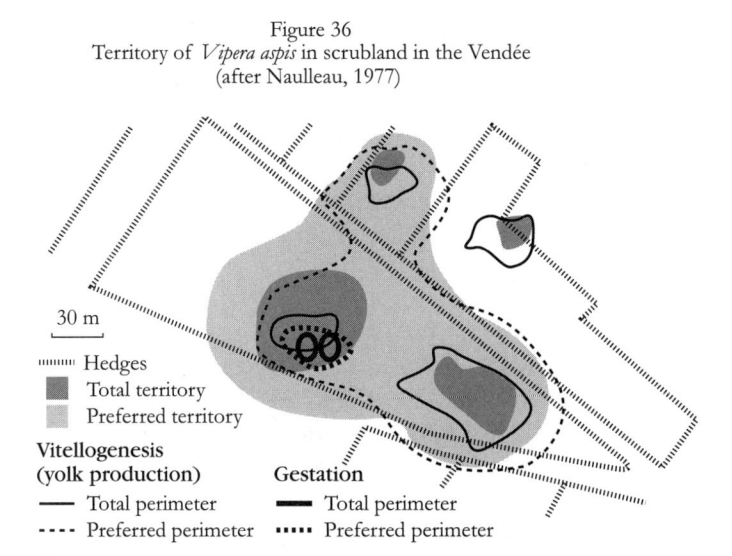

Figure 36
Territory of *Vipera aspis* in scrubland in the Vendée
(after Naulleau, 1977)

- Hedges
- Total territory
- Preferred territory

Vitellogenesis
(yolk production)
— Total perimeter
---- Preferred perimeter

Gestation
— Total perimeter
••••• Preferred perimeter

Rhabdophis tigrinus moves on average 55 m in one season and about 120 m in the following season. The maximal distance covered during a season is 345 m. *Coluber constrictor* in the United States has a territory of a dozen hectares, six times the size of that of the European viper according to the same method of calculation. However, the occupied surface varies on average from 0.4 to 12 ha depending on the region. These differences are due to food preferences (insectivorous in a continental climate and carnivorous in an oceanic climate), individual nutritional differences depending on energy needs, and prey abundance, which in turn is linked to environmental constraints. The daily movements of this species, also called "racer" are on average between 33 and 105 m according to the studies. Surface areas in the order of 10 hectares have been mentioned for some species. The territory of certain North American crotalids covers between 8 and 28 ha with an average distance covered during a year of 600 to 1,900 m. *Protobothrops flavoviridis* moves approximately 200 m a day in a maximal area of 700 m^2. Each individual returns to its starting point at the end of its circuit. The daily distance covered by

Bitis gabonica in Equatorial Africa is 25 m for the female and 50 m for the male. Essentially these movements take place at night. The territory covers between 0.8 and 1.6 ha. In addition to methodological inaccuracies, great variations can be caused by the high biomass of a snake that needs more prey, or by unfavourable environmental conditions like deserts where the scarcity of prey animals forces the snake to cover longer distances. The Pacific Ocean marine elapid *Aipysurus laevis* has a territory of 1,500 to 1,800 m² with a reef length of approximately 150 m.

Factors influencing the size of the territory are both environmental (temperature, plant cover, prey abundance and predation pressure) and physiological (energy requirements, reproduction, state of health).

1.4 Movements and emergences

Snakes move for four reasons, except in temperate countries during hibernation, which is a time of absolute rest.

- *Hunting* leaves intervals of rest of several days or weeks needed for digestion. It takes place at regular hours that differ with species, and is the main cause for the individual movements of snakes during the day-night cycle. It is limited to the territory of the snake. This activity can change with the rhythm of the seasons. Temperature, sunshine and humidity probably are the essential factors. There can be seasonal variations: in spring *Vipera aspis* prefers to hunt during the day, but in summer also sometimes at night, particularly if it is very hot during the day. In the tropical regions the times are regular and constant except for the strong rains, which incite the snakes to leave their shelter.

- *Thermoregulation* is assured in the poikilotherm snakes by an exposure to the sun, which allows the warming of the body. The length of the basking period depends on the ambient temperature, the metabolic requirements and the state of health of the snake. As it cannot by itself raise its temperature in the case of fever, it has to place itself into the sun to warm up. The excursions of the snakes also depend on the climatic conditions: sunshine, relative humidity, temperature of the air and of the substrate on which the snake moves or rests.

- *Mating* generally is a seasonal activity. The mating season does not affect the absolute population density, but the snake's search for a sexual partner increases the frequency of human/snake contacts and this is in line with a significant increase of the apparent density. In countries with a temperate climate mating takes place in spring just after the end of hibernation. In tropical regions mating generally occurs at the end of the dry season. It is to be noted that the males go in search of females. This appears to be a general phenomenon in snakes and is the reason for a very skewed sex ratio in certain seasons. In tropical areas the sex ratio can reach seven males for one female in the mating season. (Fig. 37).

Figure 37
Seasonal variation of the sex ratio of snakes in a humid tropical region (French Guyana)

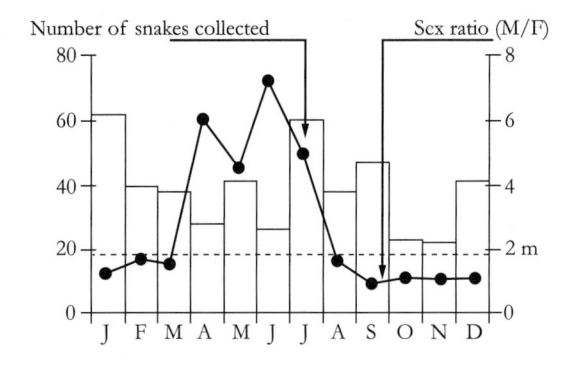

In the Languedoc the sex ratio of *Vipera aspis* varies from 0.8 to 3.7 males per female according to the season and the sexual activities of the snakes. Several species of snakes have a diphasic sexual cycle, having two reproductive seasons. Some species, particularly in the humid equatorial region, are polyestrous and reproduce throughout the year.

During gestation the females move little because of reduced nutritional requirements. In contrast they bask to regulate their temperature, which has to remain high for the maturation of the eggs.

- *Birth (hatching)* is also a seasonal activity and following it the young snakes have to go in search of their own territories. The conquest of their own territories is anarchic and takes a long time. During this period there is a multiplication of total numbers, which is seen by a very significant increase in the number of collected specimens. The number of eggs laid by the oviparous snakes and of youngsters given birth by the ovoviviparous snakes varies with the species. It can be up to 10 eggs in *Dendroaspis viridis* and *Causus maculatus*, 15 eggs in *Dendroaspis polylepis* and *Lachesis muta*, 20 hatchlings in *Trimesurus albolabris*, 20 to 25 eggs in the *Naja* spp. and *Causus rhombeatus*, 30 hatchlings in *Calloselasma rhodostoma* and *Bitis arietans*, 40 hatchlings in *Bitis gabonica*, around 50 hatchlings in *Bothrops atrox* and around 100 eggs in large females of *Python sebae*. These numbers cause a sudden increase in snake population densities, which decline rapidly during the following months due to heavy predation. It appears that there is a great diversity in the modalities of dispersion of the newborn snakes. The dispersion is immediate, continues during the first few days and can involve large distances. However, some young ones right away show a more sedentary behavior, even more so if environmental conditions induce them to further restrict their movements. For the rest, the movements appear to be dictated by chance and can take them on a circular trajectory back to their starting point after a more or less prolonged journey.

It has to be noted, though, that while egg laying is economical for the female which is freed quickly from the burden of her eggs, it is costly for the population because of the destruction of eggs. In contrast it has been shown that gestation renders females vulnerable because of an increase in predation exerted on them.

The combination of these different constraints is the reason for the seasonal variations of population densities (Fig. 38). In any case most authors have found that sexual behavior is the main factor influencing the relative population density of snakes.

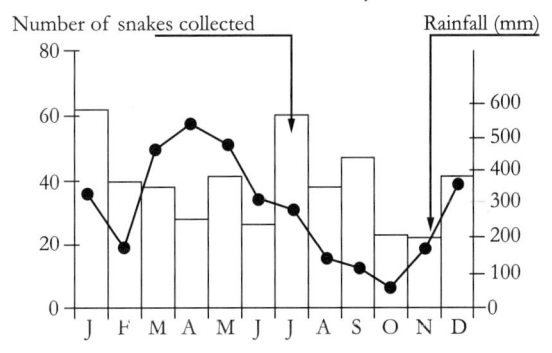

Figure 38
Seasonal variation of the abundance of snakes in French Guyana

1.5 The composition of populations

In epidemiology the composition of populations can be limited to the venomous species present in a biotope or a particular site. The identification of dangerous species allows one to anticipate the number and severity of envenomations as well as the main symptomatology to be expected. Experience has shown that there is a strict correlation between these results and morbidity or lethality.

Water and agricultural works certainly are a major factor in the redistribution of snake populations. The attraction or repulsion of certain species leads on one hand to a modification of the snake population density and on the other hand to a selection for certain species.

Each species has its own preferences which attract it to a biotope or which drive it away as a consequence of ecological changes, mostly produced by man. Schematically the snake species can be separated into three groups:

- wild species which keep away from inhabited places and tend to disappear with agricultural development; some of these are seriously threatened;

- indifferent species, the behavior of which does not appear to be much affected by human influence on the environment;

- commensal species which flourish in the immediate human environment.

Riverine or traditional village agriculture uses a small surface if one considers the respective size of each lot. The limited surface of the fields explains why there is no clear rupture with the neighboring natural environment. The snakes found there are the same as those in the adjacent bush and their behavior is very similar.

In contrast in commercial plantations the acreages used and their ecological particularities in relation to the natural environment turn them into a specific entity. Each type of plantation has a structure and a population linked to the cultivated product and to the agricultural methods used. Banana plantations are an exception as the culture methods vary from one farm to another and this leads to important differences between snake populations (Fig. 39). In some banana plantation the mulching which helps water to be retained at the foot of the trees, shel-

The Epidemiology of Envenomations

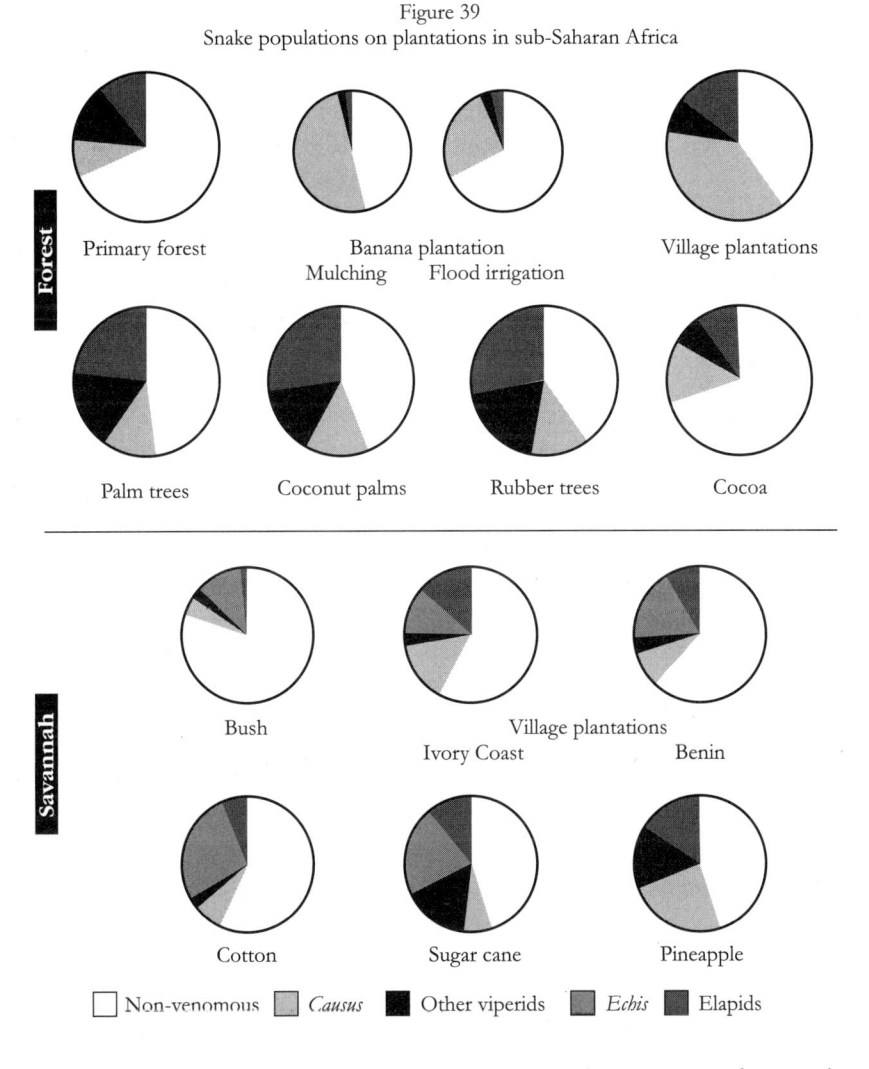

Figure 39
Snake populations on plantations in sub-Saharan Africa

Forest
- Primary forest
- Banana plantation — Mulching
- Banana plantation — Flood irrigation
- Village plantations
- Palm trees
- Coconut palms
- Rubber trees
- Cocoa

Savannah
- Bush
- Village plantations — Ivory Coast
- Village plantations — Benin
- Cotton
- Sugar cane
- Pineapple

Legend: Non-venomous | *Causus* | Other viperids | *Echis* | Elapids

ters a large population of the semi-burrowing viperid *Causus maculatus*. Other plantations that are irrigated by ditches in which water circulates between the blocks of trees, are attractive to *Afronatrix anoscopus*, a piscivorous colubrid which is aggressive but without danger to man.

In Panama the *Bothrops* spp. (crotalids with a hemorrhage-causing and necrotizing venom) constitute 25% of the snake populations on certain fruit plantations where the lethality is high. In Japan the *Protobothrops* spp. (crotalids with a mainly necrotizing and sometimes hemorrhage-causing venom) form up to 90% of the populations on the southern islands, mainly occupied by rice paddies. Venomous snakes are rare in pineapple plantations; they are represented by arboreal species in rubber tree plantations (elapids in Africa, elapids and crotalids in Asia), by mammal-eating snakes in palm plantations and aquatic ones in rice fields.

In urban zones there is a much stricter selection of species. However, some venomous snakes show a disquieting commensalism with man (*Bothrops atrox* in South America, *Naja nigricollis* in intertropical Africa and certain species of crotalids in South East Asia). While the snake population is thin, the human population there is very high and consequently there is a considerable risk of encounters.

Many venomous species are ubiquitous and commensals of man. In Europe *Vipera aspis* adapts remarkably to environmental transformation, though without leaving the strictly rural environment. In contrast in tropical regions *Causus maculatus*, *Naja nigricollis* and to a lesser degree *Echis ocellatus*, *Naja melanoleuca*, *Dendroaspis jamesoni* and *D. viridis* stay in the urban environment with a preference for rubbish dumps, where their preferred prey abounds.

2. Human Behavior

2.1 Agricultural and pastoral activities

Human activities in the rural environment represent a real exposure to snakebites. However, important variations are seen as function of the biotope. This is due to the behavior of the snakes as well as to the agricultural practices.

In the industrialized countries a small proportion of the population is engaged in agriculture, which in addition is heavily mechanized. Consequently the risk of

The Epidemiology of Envenomations

snakebites is reduced. Even in industrialized countries that have many species dangerous to man, the incidence remains low. However, tourism and recreational activities in contact with more or less managed nature are growing steadily and this increases the relative risk of accidents.

If in temperate countries snakebites are rare events, their incidence can be considerable in tropical or equatorial countries.

Non- or poorly mechanized agricultural activities are the source of the largest number of accidents. It is possible to quantify more precisely the risk of exposure by measuring the time taken for the work and its nature. This is relatively easy in a commercial plantation where the number of workers and their work are programmed and measured precisely. It is more difficult in subsistence plantations for which this type of information is not always available.

Human occupations are also seasonal, particularly in the plantations and this shows that the distribution of snakebites due to human/snake encounters is not dictated by chance.

In some cases the agricultural work program leads to a paradoxical situation where the number of bites does not correspond to the observed population density of the snakes. Some agricultural activities, even if only short, lead more than others to a contact between workers and snakes. Only a close surveillance of the apparent density allows the identification of the unforeseen event that leads to the paradoxical situation. The manual cleaning of the irrigation ditches in banana plantation is a perfect illustration of this phenomenon. It takes place outside the seasons of intensive activities (of man as well as of snakes) and during a short time span and with a small number of personnel, yet it still represents a particularly high risk.

The number of people cultivating at any one time in traditional village plantations is small, meaning a low density of workers with variable working hours.

In large commercial plantations human density is high and often the workers are engaged in a collective activity. Regular hours tend to uniform the risk. There the incidence is generally higher than in village plantations. The morbidity varies with the species composition of the snake population and this shows the importance of its identification. In some banana plantations a ten times higher incidence is registered than in adjoining subsistence plantations. However, the lethality there is much lower. In any case, even though half of the specimens collected are *Causus maculatus*, a small and relatively aggressive viperid, the risk of severe or lethal complications is low because of the weak toxicity of the venom. Lethality is limited on the plantations in certain developed or emergent countries (Japan, Central America) because of a health service specially designed for combating the risks of snakebite. Functional sequelae nonetheless affect 10% of the bitten persons because of the particular character of certain venoms.

2.2 Other activities

Some activities favor contact with snakes: hunting, fishing and camping offer equal occasions for an encounter. The snakes are hidden under stones, under heaps of dead wood, in the bushes and can attack when surprised or when they feel cornered.

Amongst tourists and expatriates the incidence of envenomations probably is less than 10 per 100,000 persons a year. This is because their activities do not lead to contact with venomous animals and perhaps also because of a certain prudence in their behavior or due to appropriate protective clothing. In contrast the risk linked to certain types of behavior has to be underlined. In some tropical countries walking barefoot even in the house, sticking a hand into a hole in a tree or rock or termite mount can have dire consequences.

Much attention is also given to the risk of envenomations in armies called to operate in zones populated by venomous snakes. Whatever the incidence – of 10 to 50 bites per 100,000 soldiers operating in tropical regions – the main preoccupation of headquarters on one hand is to have in place the specific means to treat the envenomations correctly (appropriate anti-venoms in sufficient quantity) and on the other hand to avoid an outbreak of panic amongst the troupes facing a risk for which they are not prepared. An American study showed that in Vietnam the fear of snakes predominated above all other fears suffered by the soldiers during the campaign. According to official sources the American army lost 5 to 10 soldiers per year during the course of the Vietnam war as a consequence of envenomations, which is negligible in view of the number of deaths in combat or even due to traffic accidents or perhaps suicides.

2.3 Handling snakes

In contrast and particularly in industrialized countries a new type of bites has been developing over the last three decades. In the Anglo-Saxon literature they are called illegitimate bites and hazardous bites in the first Francophone publications ("morsures hasardeuses"); but the term induced bites appears to be more appropriate ("morsures induites"). They happen to professional or amateur herpetologists during the intentional handling of snakes. The development of the phenomenon of the new companion animals (NCA) amongst which reptiles are well represented, leads to the prediction of an increase and expansion of this risk, also in emerging countries. Induced bites have remarkable particularities: the aggressor is identified, often there is an intensive care unit nearby, at least in industrialized countries, and access to therapeutic information pertinent to the field of envenomations is simple and rapid. However, severe clinical cases are observed: generally the snakes kept in captivity are dangerous species; in addition they are well nourished and inclined to inoculate more venom as they are being handled with little possibility of fleeing.

3. The Epidemiology of Bites

Our knowledge of the epidemiology of snakebites is very patchy. It is likely that in the developing countries where they occur most frequently, the victims of snakebite envenomations do not reach a health care center for a variety of reasons. Also the health statistics do not distinguish in these countries between snakebites and other accidents or intoxications. Also the reporting of cases of envenomation is not obligatory except in a few countries (Myanmar for example). Although one cannot speak of a deficient system of health information, the lack of precision does not allow the adoption of measures directed against this public health problem.

3.1 Study methods

The basic terminology is not universally used but is encountered more and more in recent publications.

- *Incidence:* This is the total number of bites reported per 100,000 inhabitants per annum irrespective of their severity. This figure allows the identification of the need for services in a region or in a particular sector of activity. In this latter case it will not be expressed per 100,000 inhabitants but per 100,000 people at risk within a studied sector of activity (agriculture, students, soldiers in a campaign, tourists etc.). The incidence is total if all the accidents have been reported in an exhaustive fashion. It is apparent or relative if the registration of accidents has been carried out only partially. One can for example consider the bites registered by the public health system and leave out the cases presented to other health centers (private or traditional) or those not presented because of being too far away from any health care center or because death occurred during transit. Certainly the ideal is to count all the events (total incidence) and if necessary evaluate them in one equation.

- *Morbidity:* This is the total number of bites causing clinical problems which need medical care per 100,000 inhabitants or persons concerned by the enquiry per annum. This index is important for measuring the severity of the accidents and to define the health needs of a region. The morbidity is easier to establish than the incidence. In effect the practice shows that envenomations are reported more often than asymptomatic bites, which often escape in a modern health care system or are not registered by it. The advent of severe clinical problems often leads the victim or his family to demand assistance from modern medicine even if at first traditional medicine had been solicited. The geographical distance from an official health care center is a parameter that now becomes of great importance. If the symptoms are indistinct the victim may be inclined not to consult such a service. Inversely a severe envenomation can cause the death of the victim before his/her arrival at the dispensary or hospital. It is important to note the delay between the bite and the arrival at the health care center. It quantifies not only the distance between the place of accident and the hospital, but also the motivation of the victim to come for a consultation and finally his adherence to the health care system. The comparison between incidence and morbidity defines the level of severity of snakebites in a given region. In relation to this level the mean delay of consultation allows one to evaluate the confidence accorded by the population to the care system offered and the efficacy of the means of evacuation to the latter. In tropical zones a delay of less than two hours is very good and points at the correct use of the health system. In excess of six hours it shows insufficient means of transport or a lack of confidence in the health care system. It is necessary also to record the precise symptoms observed in the patient. In practice it is most often sufficient to record the proportion of the major syndromes observed (neurotoxic, hemorrhagic, inflammatory, necrotic).

- *Lethality:* This is the percentage of deaths due to envenomation out of the total number of snakebites. It is a sensitive indicator of the severity of snakebites in the region studied. The lethality includes many factors related to the snake species encountered, to the activity of the population and to the offered health services. It is necessary to relate the lethality to the total incidence in order to avoid a bias through the involuntary selection of victims. Most often it is calculated on the basis of hospital statistics and thus the lethality is arbitrarily exaggerated, loading the real risk of snakebites. The number of deaths generally is easy to establish from official sources. The hospital registers have to be consulted first of all, but one has to take into account the deaths that occurred outside the medical system, which happens often in tropical countries. The mean consultation delay can serve as an indicator: the higher it is, the higher is the risk of finding deaths not registered by the medical statistics.

- *Mortality:* This is a current medical index. It is the number of annual deaths due to snakebite in relation to 100,000 inhabitants. The information does not provide any data on the circumstances of the snakebites nor on their severity. It allows one to establish geographical comparisons of the snakebite risks and of the health care measures which are provided as response. The mortality also allows one to verify the figures of incidence and lethality, provided they were obtained by different and independent means. The mortality equals the product of incidence times lethality.

3.2 Circumstances of the bites

Snakes only bite to defend themselves and to protect their escape. No species is aggressive in the sense that it would deliberately attack people. In addition the inoculation of the venom is a voluntary act. It is therefore not an unavoidable phenomenon but the response to a critical situation.

The bite therefore is the direct consequence of accidental or intentional approaching between man and snake. There is a distinction between the dry bite without penetration of the venom, the envenomation, which results from the pharmacological action of the venom and the reaction of the body that is triggered by it.

3.2.1 Activities at the time of the bite

In industrialized countries practically all bites happen during recreational activities, when the victim disturbs the snake or steps on it. Accidents of an occupational nature involving farmers, foresters or road workers have become the exception, either because of the heavy mechanization of their activities or because they wear protective clothing against bites. In contrast hikers, sportspeople and strollers are dressed in much lighter clothes and have a more careless attitude towards the risk posed by snakebite. They expose themselves by lying down in the grass, bringing their hands close to a potential shelter like a heap of wood or stones, a rock cleft, a declivity, a hedge, or even by sitting down on pack wall.

In tropical regions, particularly in developing countries, the accidents are linked much more often to occupational activities. Three quarters of the bites happen during agricultural work, hunting or while walking to or from work.

Traditional methods of agriculture lead to a serious exposure, notably when the hands are close to the ground or the tools are primitive and short.

3.2.2 Induced bites

The geographical distribution of induced bites is the opposite of that of haphazard or accidental bites. They are mainly observed in industrialized regions with a high urban density. The annual incidence is low. In France it is approximately one or two cases per million inhabitants per year. In the United States the induced bites form 20% of the envenomations treated in hospitals. This phenomenon tends to become general, even in some emerging countries. In the Johannesburg region 60% of bites treated in hospitals are induced bites.

There is no seasonal or day/night periodicity. The accidents happen almost exclusively to adults, most often young adults, prevalently males. In more than 90% of the cases the bite occurs on the hands.

The Epidemiology of Envenomations

The risk most often involves the snake keeper rather than a visitor or neighbor and depends on two factors: the number of snakes kept and the time of exposure of the handler. The relative risk increases clearly up to a number of about 50 snakes kept and then decreases and becomes stable beyond 1,000 snakes. The experience acquired by the handler reduces the risk significantly. The relative frequency of bites is around one accident every five years irrespective of the experience of the handler and the type of manipulation. Contrary to a widespread opinion hobbyists are not more exposed than professionals, doubtless because the latter handle venomous snakes more often and longer at each session thereby proportionately increasing their exposure. Perhaps also their attention decreases involuntarily during prolonged handling. The simple keeping of snakes does not *a priori* cause great risks. The dangerous activities are cleaning the cages without bothering to remove the occupant, veterinary care, and venom collection. Most of these can be carried out under safe conditions, if the snake is handled with the help of instruments or after it has been anaesthetized. However, certain acts that are also known to be dangerous lead to a similar frequency of accidents, probably because of their trivialization or the inattention of the handler.

The morbidity is higher than 55% of the bites. Complications, notably necroses, occur frequently, at least in more than 15% of the cases. The lethality remains high in spite of the precautions that can be taken: early and quick evacuation to an experienced medical center, availability of suitable therapeutic means. The probable reason is that the snakes responsible for the accidents belong to species with strongly toxic venom and that the individuals often are of large size. In addition handling the snake and keeping it in captivity can provoke in it a stronger aggressive reaction leading to the injection of a larger quantity of venom.

3.3 Characteristics of bites

Apart from induced bites that mainly affect men, the incidence is equal in both sexes in industrialized countries. There is also no particular age of exposure.

In temperate countries the bites occur between spring and autumn, mainly during the day. There is an increase during the summer vacations.

A large proportion of the bites, between 50 and 70% depending on the country, occurs on the lower extremities. The hands are involved in a quarter or third of the accidents, head and body in the other cases.

In developing countries young men are most often involved: they receive between 50 and 75% of the bites. Although children represent close to half of the general population, they are rarely bitten, as are women. Yet these latter are engaged in activities with a risk similar or even higher than that of men.

The seasonal incidence of accidents is linked to the behaviour of the snakes and to the agricultural calendar (Fig. 40).

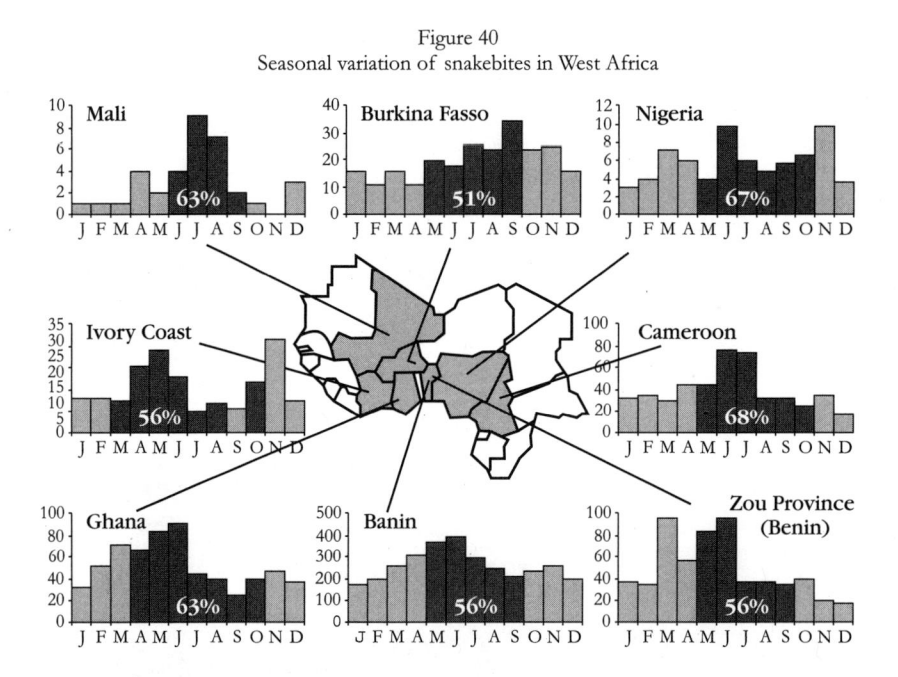

Figure 40
Seasonal variation of snakebites in West Africa

There are some geographical variations linked to agricultural practices: in forest regions the bites are more spread out throughout the year, while in the savannah the accidents occur more frequently in the rainy season (Fig. 41). The relation with rainfall figures shows its direct effect on human and snake behavior (Fig. 42).

The majority of bites occurs during late afternoon or early evening. Some take place during the night in the house and are inflicted during sleep by snakes moving in the house in search of prey.

More than 80% of the bites are on the lower limbs, mainly below the knee, but large geographical variations are seen. Bites of the hand are rare without being exceptional, notably in peasants working with tools with a short handle or in children while exploring with naked hands burrows in search of small vertebrates to add to their nutrition.

3.4 Severity of bites

The severity of bites is influenced by several factors. The toxicity of the venom and the quantity injected by the snake evidently are essential elements. These factors depend on the species of snake, its size, the capacity of its venom glands, their state of filling and the circumstances of the bite.

Age, size and state of health of the victim and the site of the bite are equally important elements.

Figure 41
Seasonal variation of snakebites in relation to rainfall

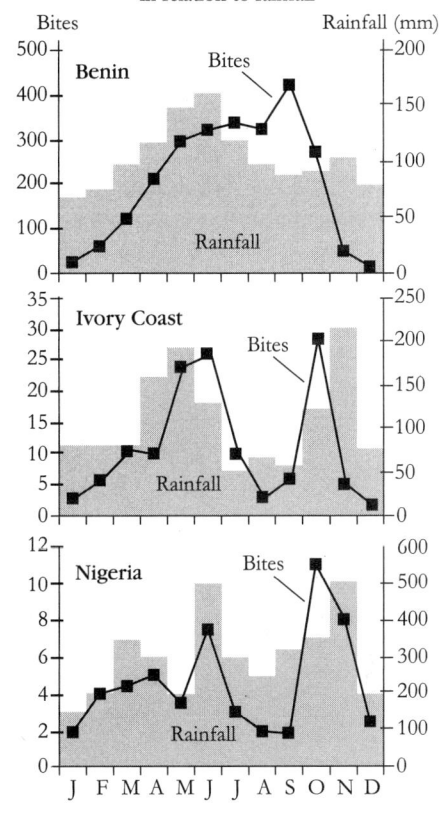

Figure 42
Seasonal variation of envenomations in the south of Japan (after Sawai et al. *in* Rodda *et al.*, 1999)

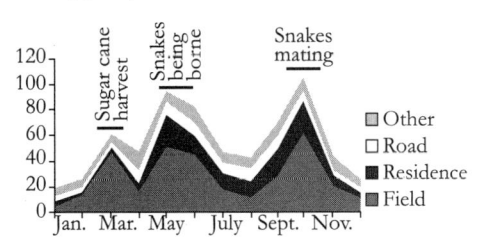

The delay before a consultation will also have important consequences. A delay is a source of complications and diminishes the efficacy of the treatment in proportions that are difficult to evaluate.

The prognostic factors of an amputation have been studied in Brazil in almost 3,000 patients bitten by *Bothrops* spp. Twenty-one patients (7%) underwent an amputation. A significant correlation was noted between the risks of amputation on one hand and on the other hand the month of the accident, the hour at which the bite occurred, the size of the snake, the anatomical site of the bite, the presence of systemic bleeding and renal insufficiency. The patients bitten in a finger, during the cool months, between 0 h and noon, by a snake longer than 60 cm, and who presented blood blisters, abscesses at the site of the bite, systemic bleeding and/or renal insufficiency, have a significantly higher risk than the other patients. These different characteristics are due to the behavior of a particular species, the venom of which is more toxic and leads to this type of complications.

Care rendered defectively due to the lack of health structures or the absence of appropriate materials and medicines increases the risk of an unfavorable course irrespective of the delay of consultation. The first aid given, if it is aggressive – tourniquet, incisions, scarification, septic cataplasm or other – brings the danger of reducing the circulation, infecting the wounds or provoking hemorrhages. A secondary infection aggravates the local lesions and leads to invalidating sequels. In a study in Thailand iatrogenic complications linked notably to prolonged artificial respiration in cobra-type envenomations, occurred in 25% of the patients. Local secondary infections and tetanus affected 15% of patients suffering from envenomation. Amongst other causes a therapeutic error, failure of the artificial respiration or its blockage due to incompetent or overworked medical personnel, or lack of appropriate therapeutic means or the absence of antivenin were incriminated in more than half of the deaths. A delay in consultation through negligence, through inaccessibility of health care centers or delayed because of first seeking an occidental treatment, was directly responsible for 20% of the deaths. In Africa the lethality could be reduced by 90% if a correct treatment could be instituted in time.

The frequency and characteristics of dry bites, those that are asymptomatic because inflicted by a non-venomous snake or by a venomous one which did not inject its venom, have been studied in many countries: United States, Brazil, India, Thailand, Japan, Cameroon, Ivory Coast, South Africa and Kenya. Depending on the biotopes and the credibility of the medical statistics of the country, they represent 20 to 65% of the bites. Even in France the incidence of dry bites is very variable. Globally it is considered to be close to 50%. A more precise study carried out in the Yonne, a rural department south east of Paris, has found this incidence to be less than 20% of the patients. In Brazil and Thailand it rises to 40% of bites and in West Africa to 40% in the savannah and 60% in the forest. In Gabon and Kenya the prevalence of dry bites reaches 80%. The epidemiology differs between the populations with dry bites and those that suffer from envenomations. The sex ratio and the hour are identical, but the age of the victims, the site of the bite and the season are not the same in any given geographical region. This is due probably to behavioral differences of the snakes responsible for the bites.

3.5 Geographical distribution of envenomations

The real frequency of envenomations throughout the world and their severity remain largely unknown, with the exception of some countries where these events are rare or correctly reported (Tables XIII and XIV, Fig. 43). This information is essential to determine the steps to be taken in cases of bites, to foresee which stocks of medicines are needed, notably antivenom sera, and to define the course of medical treatment.

Table XIII
Incidence and severity of snake bites world-wide

region	population (x 10^6)	number of bites	number of envenomations	number of deaths
Europe	750	25 000	8 000	30
Middle East	160	20 000	15 000	100
USA – Canada	270	45 000	6 500	15
Latin America	400	300 000	150 000	5 000
Africa	750	1 000 000	500 000	20 000
Asia	3 000	4 000 000	2 000 000	100 000
Oceania	20*	10 000	3 000	200
Total	5 100	5 400 000	2 682 500	125 345

* population of the regions with terrestrial venomous snakes

Table XIV
Incidence and severity of snake bites in the most exposed countries
(the numbers in brackets correspond to recent figures)

country	incidence (bites /100 000 h.)	morbidity (envenomations /100 000 h.)	lethality in hospitals (% deaths)	mortality (deaths/ 100 000 h.)
Spain	5,5	2	0,8	0,02
Finland		3,8	0,5*	0,02
France	3,5	0,4	0,2	0,00
Great-Britain	0,7	0,3	0,01	10^{-4}
Italy	2,5	1	0,2	0,002
Poland		0,3	0,3	0,001
Sweden		2,6	0,5*	0,01*
Switzerland	0,6	0,14	0,1	0,001
Saudi Arabia		10		
Israël		7	1,3	0,03
Jordania		0,36	6,2	0,02
Canada	1	0,8	0,01	10^{-4}
United States	10	2,5	0,3	0,005
Mexico		1	0,7	0,007
Brazil		15	1	0,1
Colombia		10	5	1,6
Costa Rica		30	3	0,4
Ecuador	125	30	5	1,5
French Guyana	125	75	2,2	1,5
South Africa		(35)	(5)	(0,5)
Benin	200-650	20-450	3-15	10,1
Burkina Faso	(600)	(7,5-120)	(1,9-11,7)	(0,3-2,4)
Cameroon		(75-200)	(5-10)	
Congo	(12-430)	(20)	(1-6,6)	
Côte d'Ivoire		100-400	2-5	
Gabon		(20-90)	(2,5)	

Table XIV *(continued)*

country	incidence (bites /100 000 h.)	morbidity (envenomations /100 000 h.)	lethality in hospitals (% deaths)	mortality (deaths/ 100 000 h.)
Gambia			14**	
Guinea		(100-150)	(18)	
Kenya	(150)	(5-70)	(2,6-9,4)	(0,5-6,7)
Mali		(90)	(6,8-17,6)	(15,8)
Natal (South Africa)		100		
Niger			(5,1-6,9)	
Nigeria	160 (48-603)	(100-150)	(2,1-27)	15,6
Senegal	(15-195)	(20-150)	(7)	(1,5-14)
Togo		130	3	4,5
Zimbabwe		3,5	(1,8-5)	
Korea			5	
India	(65-165)	(1,4-6,8)	(1,7-20)	(1,1-2,5)
Indonesia			(2,5)	
Japan	1-580	0,2-580	0,7	0,01
Malaysia	(400-650)	(130-300)	(0,4-1)	0,2
Myanmar	35-200	35	10	3,3
Nepal		110	3,8-5,2	5,6
Sri Lanka	(250)	(45-60)	(4-22)	6
Taiwan	30-300	80	2,1-2,7	1,6-2,7
Thailand		10	0,9	0,05
Australia	3-18		0,1-0,4	0,03
Papua New Guinea	215-530	3-85	1,9,6	7,9

* before the present use of the serum therapy
** children

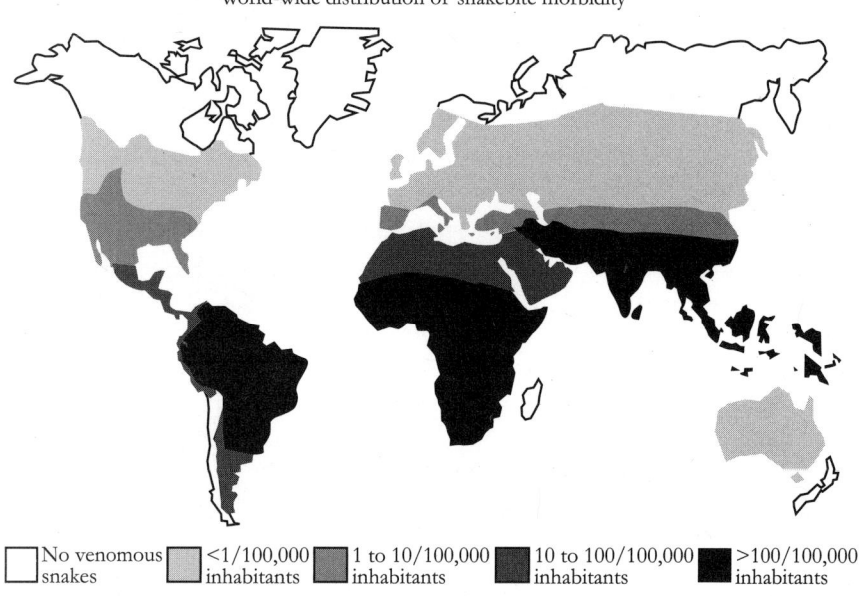

Figure 43
World-wide distribution of snakebite morbidity

☐ No venomous snakes ▧ <1/100,000 inhabitants ▨ 1 to 10/100,000 inhabitants ▩ 10 to 100/100,000 inhabitants ■ >100/100,000 inhabitants

3.5.1 Europe

In Europe snakebites are relatively rare. The snakes responsible for accidents belong to the genus *Vipera*, which in Europe are represented by some species that are not among the most venomous ones: *V. ammodytes, V. aspis, V. berus* and *V. latastei* and their subspecies.

In France the mean incidence has been estimated to be around 3.5 per 100,000 inhabitants (Fig. 44). This corresponds to approximately 2,000 bites or close to 500 envenomations and one death per year.

There are around 50 induced bites per annum causing one death every five years. In the Department of the Yonne, about 150 km southeast of Paris, an annual incidence of 5 cases per 100,000 inhabitants has been observed. These cases certainly are not distributed homogenously in the Department, but the incidence appears to be largely representative of envenomation accidents in rural regions of Western

The Epidemiology of Envenomations

Europe. Studies carried out in Vendée, in the Limousin, the Poitou and the Massif Central corroborate the results from the Yonne. Men are bitten more often (61% of the accidents) than women. Children below 16 years are involved in 40% of the cases, but persons of 60 years and over only in 10%. Most of the cases occur in summer, between May and September. In this period fall 90% of the cases with a peak in July. Most of these cases are treated by general medical practitioners, although lately there has been a net increase in the number of cases treated in hospitals.

Figure 44
Distribution of the incidence of snakebites in Metropolitan France

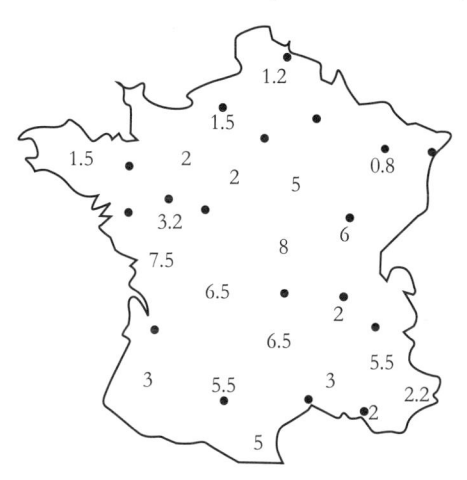

• Anti-poison center
The numbers indicate the annual regional incidence

In Great Britain there are about 200 hospitalizations per annum due to snakebites, but no deaths have been reported since 1975. More than half of the bites affect the hands, which probably is in relation to high frequency of induced bites.

In Switzerland the morbidity is low, approximately some tens of envenomations per annum.

In Spain the annual incidence can pass above 2,000 bites including 3 to 7 deaths per year.

In Italy the number of bites lies between 1,500 and 2,000, with 800 envenomations annually and almost one death on average. Induced bites appear to be relatively rare in southern Europe (Spain, Portugal, Italy, Greece).

No precise numbers are available from Germany, but it is known that induced bites occur very frequently there.

In Poland the number of envenomations is estimated to be on average 150 per year with a death occurring every two or three years.

In Sweden there are about 200 to 250 envenomation per annum. The lethality has become practically zero since the use of anti-venom while it was 0.5% before its use.

In Finland where 150 to 200 bites are recorded each year, the mortality was 0.5% before the 1960s, but has been reduced considerably since the regular use of antivenom serum. More than 70% of bites are treated by serum-therapy and about one third of the cases are hospitalized.

In a population of 750 million inhabitants the number of snake bites in Europe reaches 25,000 cases, of which 8,000 present a patent envenomation. Most victims are hospitalized and receive an appropriate treatment. Some 30 deaths occur each year, mostly involving induced bites by exotic venomous snakes.

3.5.2 Near East and Middle East

In the Near East there are more, more diversified and more dangerous venomous species than in Europe. The main ones are *Macrovipera lebetina*, *Daboia palaestinae* and *Montivipera xanthina*; also there is *Atractaspis enggadensis*. The elapids (*Naja haje* and *Walterinnesia aegyptica*) rarely are responsible for bites. The species of the genera *Cerastes* and *Pseudocerastes* – the former are relatively common in the south and east of the Mediterranian Region – appear to be relatively less dangerous. The epidemiological data concerning snakebites remain very patchy.

In Israel 200 to 300 envenomations are reported annually; the morbidity is ten times higher in rural cultivated zones. The mortality is 0.03 per 100,000 inhabitants per year. Most bites occur on the lower extremities. The lethality is low because of early and correct treatment.

In Saudi Arabia 85% of the bites are observed in the hot season and 60% affect adults. The number of envenomations is estimated at 2,500 per annum.

In Jordan around 20 patients with snakebite envenomation are admitted to hospitals. The lethality of these cases is 6.2%

The Epidemiology of Envenomations

As in North Africa it is likely that scorpion stings occur much more frequently. In a population of around 160 million inhabitants the number of snakebites can be more than 20,000 per annum with 15,000 envenomations and around 100 deaths. Probably the frequency of hospitalization of the victims is lower than in Europe.

3.5.3 The Americas

In Canada and the United States the incidence of snakebites is not much different from that observed in Europe. More than 45,000 snakebites happen each year in North America of which approximately 10,000 are due to venomous species and 6,500 need medical interventions. People from 15 to 30 years make up 50% of the patients. In 45% of the cases the bite is inflicted during an agricultural activity and 60% affect the feet. More than 90% happen during the day between 6 and 20 h. The number of induced bites there is one of the highest in the world. The lethality is low if one considers the high toxicity of the venom of the crotalid species that are encountered in those regions. The agricultural states in the south have a higher mortality than those in the north, particularly in the industrialized states (Fig. 45).

Accordingly less than 15 deaths are recorded each year. Also these deaths are generally caused by a delayed start of the treatment or, in certain communities, by a refusal to undergo medical treatment.

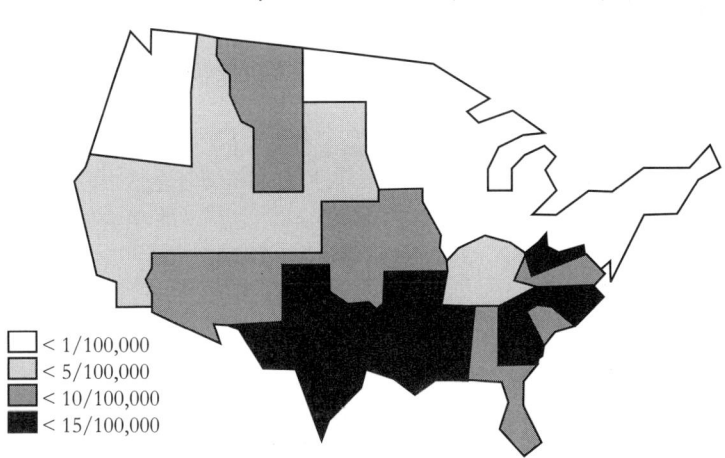

Figure 45
Snakebite morbidity in the United States (after Parish, 1980)

In Central and South America the prevalence of snakebites is significantly higher. The crotalids cause most of the envenomations with notably the genus *Bothrops* in all of South America, followed by *Crotalus durissus durissus* in Central America and *Crotalus durissus terrificus* in South America. The use of antivenom sera has clearly improved the prognostics of the envenomations. The particular effort to develop antivenoms with better tolerance and higher efficacy since about 10 years ago needs to be mentioned here.

In Mexico there are about 1,000 envenomations per year. This figure is influenced on one hand by a strong urbanization and on the other hand by certain under-estimation of this phenomenon. The morbidity is three to four times higher in rural zones, but still low. Less than 10 deaths are reported every year, which appears to be very much under-estimated. One third of the bites happen in summer. Almost half of the victims are peasants bitten during agricultural work. Three quarters of the bites affect the lower members. Crotalids are responsible for 45% of the envenomations and the *Bothrops* spp. for 43%. Mexico is a zone of high endemicity of scorpions with a morbidity of 65 envenomations per 100,000 inhabitants per year.

In Ecuador close to 4,000 cases of envenomations are hospitalized each year and around 200 of these die. In the forest regions the incidence can reach 1,000 bites for 100,000 Indians (Native Americans) per year.

In Costa Rica a little over 1,000 envenomations are declared of which 30 to 40 die. The highest incidence is observed in the high-rainfall and agricultural regions. Close to one quarter of the deaths occurs in patients between 10 and 19 years. Young men engaged in agricultural activities suffer most of the envenomations. The mortality has declined from 5 in 100,000 inhabitants in 1950 to less than 1 in 100,000 in 1990, because of more frequent and earlier hospitalization and an improvement of the antivenoms. In Columbia about 4,000 envenomations are counted per year for 100,000 inhabitants in rural areas with 200 deaths and 250 patients suffering sequelae, mostly necroses. *Bothrops* spp. are incriminated in 95% of the cases, notably *B. asper*.

In Brazil 25,000 envenomations are reported each year for the whole country, but the incidence in the state of Amazonia lies between 350 and 450 bites per 100,000

The Epidemiology of Envenomations

inhabitants. The morbidity can reach 200 envenomations per 100,000 inhabitants a year. The mortality lies at 0.4 death per 100,000 inhabitants in the south of the country against 5 per 100,000 inhabitants in some regions in the north of Brazil. Equally the lethality is below 1% in the south of the country (0.3% in Sao Paulo, 0.7% in the Ceara) but can go up to 6.5% in northern forests. In Bahia the incidence varies between 25 and 120 bites per 100,000 inhabitants, of which 60% occur in the rural zone. Peasants represent 65% of the victims. In Belo Horizonte the frequency of bites is 15 times higher in the country than in the urban zone, and 7 times higher during agricultural work than for any other activities.

In French Guyana 110 to 120 envenomations are hospitalized every year with 2 or 3 deaths. The incidence lies between 45 and 590 bites per 100,000 inhabitants per year depending on the region.

The total incidence of snakebites in Latin America is estimated at 300,000 accidents for 400 million inhabitants. Half of these are envenomations that need an appropriate treatment, of which probably only two thirds of the victims benefit in reality. The annual number of deaths due to envenomations probably surpasses 5,000.

3.5.4 Africa

In North Africa the morbidity is close to 15 envenomations due to snakebite per 100,000 inhabitants which is negligible in comparison to scorpion stings which amount to 550 per 100,000 inhabitants per year. Most envenomations are caused by viperids (*Macrovipera lebetina, Echis pyramidum, Cerastes cerastes* and *C. vipera*) while the bites of *Naja haje* and *Walterinnesia aegyptica* are the exception.

In Sub-Saharan Africa the prevalence is strongly under-estimated by the medical authorities. This has two reasons: On one hand the case reporting system is not appropriate and on the other hand a large proportion of the victims do not go to the modern health care centers. Most of the epidemiological studies are focal and in view of the heterogeny of situations observed on the continent do not allow an overall evaluation of the morbidity.

Many venomous species are found in Africa. Amongst the viperids *Echis ocellatus* and *E. leucogaster* have to be mentioned which cause close to 90% of the envenomations in the Sahel and the savannah zones. The vipers of the genus *Bitis*, in the savannah

B. arietans and in the forest *B. nasicornis* and *B. gabonica*, are incriminated in a small number of accidents, which always are serious, though. In the forest as in the savannah *Causus maculatus* is a small and not very dangerous viper, but often involved because of its aggressiveness and its ubiquity. The elapids belonging to the genera *Naja* and *Dendroaspis*, the former ones in the savannah as well as in the forest and the latter ones more in the forests, they produce fewer bites, but these are generally severe and need an early treatment. In the savannah zone during the rainy season 10 to 20% of all hospitalized patients are admitted because of snakebites.

In Senegal there is a great disparity of incidence in the different regions. In the southern forest zone the incidence is probably above 200 bites per 100,000 inhabitants per year. In the Sahelian regions, particularly in densely populated ones, the annual incidence is around 25 bites per 100,000 inhabitants. The lethality is high everywhere, around 7%. In rural zones traditional medicine is consulted prevalently. In Sine-Saloum 95% of the victims consult a traditional healer first or instead of a medical dispensary. The mortality can be high in some places: in the south, in Casamance it reaches 14 per 100,000 inhabitants.

In the Gambia, where the morbidity is doubtless close to that in Casamance, the lethality in children is 14%. The mean delay before consultation is over 10 hours after the bite.

In Guinea the morbidity is between 100 and 150 envenomations per 100,000 inhabitants per year, with a lethality of 18% and 2% amputations. The incidence certainly is under-evaluated. In the southern forest region the cobra syndrome is seen in 35% of the cases and the inflammatory or viper syndrome in close to 30%.

In the rubber plantations in Liberia the annual incidence is 420 bites per 100,000 laborers and the morbidity 170. However, the lethality is below 1%.

In the south of the Ivory Coast the annual morbidity lies between 200 and 400 envenomations per 100,000 inhabitants with a lethality of 2%. In the Lagune region the morbidity lies close to 100 envenomations per 100,000 inhabitants per year; in the forest the morbidity is above 200 envenomations per annum and the same applies to the mountainous region. More than 80% of the victims are 15 to 50 years old. In the north of the country, in the savannah, the morbidity lies at

230 envenomations per 100,000 inhabitants per year with lethality of 3 to 5%. Agricultural laborers are particularly affected, notably in the industrial plantations, where 25% of the bites occur in 1.5% of the population. The incidence can reach 3,000 bites per 100,000 workers, ten times more than in the neighboring subsistence plantations.

In the south of Ghana the morbidity is 20 envenomations per 100,000 inhabitants with a lethality of 0.2%. These figures are relatively old and certainly represent an under-evaluation of the real number of cases. The towns are not free of this phenomenon as even the largest ones have an incidence of 10 bites per 100,000 inhabitants per year.

In Togo the health services record an average of 5,000 cases per annum with 150 to 200 deaths.

In Benin 4,500 envenomations are recorded by the public health services with more than 650 deaths. There is a great geographical disparity. In the South the morbidity lies between 20 and 450 envenomations per 100,000 inhabitants per year with a lethality of 3% to 10%. In the north the incidence lies between 210 and 650 cases per 100,000 inhabitants with a morbidity as high as 300 envenomations per 100,000 inhabitants and a lethality of close to 5%. In the industrial sugar plantations the incidence is 1,300 bites in 100,000 workers with a lethality below 1.5%. In the rural environment the annual incidence is 430 bites per 100,000 inhabitants, including dry bites. Most of the bites occur during the rainy season. Less than one third of patients bitten by a snake consult a hospital, 80% of the patients first seek the help of traditional medicine. The population at risk consists mainly of active men. Men receive twice as many bites than women and over 60% of the patients are between 21 and 50 years old. The frequency of inflammatory syndromes is 66%, that of hemorrhagic syndromes 12% and necroses occur in 5% of the cases.

In Mali the morbidity is close to 100. The mean lethality is 7% of hospitalized cases with peaks in certain regions.

In Burkina Faso the morbidity lies over 100 cases per 100,000 inhabitants with a lethality of 2 to 12% depending on the kind of therapy applied. The morbidity in Ouagadougou, the capital of Burkina Faso with a population of approximately 750,000, lies at 7.5 envenomations per 100,000 inhabitants and most of the cases occur in the periphery of the town, in a still poorly and irregularly serviced zone. In the rural zones the incidence is much higher. The sex ratio of patients is 1.5 male to 1 female. More than 40% of the victims are over 15 years old. Studies carried out in the Ouagadougou region (Mossi area) revealed that the incidence could reach 500, even 700 bites per 100,000 inhabitants. The morbidity reported by the health system lies between 35 and 120 envenomations per 100,000 inhabitants, which translates into 7,000 and 10,000 envenomations per year causing 200 deaths as reported by the national health services. Several studies on treatments sought have shown that less than a third of the bite victims consulted health care centers. From this can concluded that the health statistics record only 35% of the envenomations and scarcely 20% of the deaths. Also, half of those would happen before any therapeutic intervention. The average delay separating the bite from the consultation in a health care center exceeds in effect 50 hours. In spite of this more than 15% of hospital beds are occupied by snake envenomation victims during the rainy season.

In Niger the incidence appears to be relatively low, probably because of low human as well as snake population densities that limit the encounters. However, the lethality is high because of the very low rate of consulting modern health care centers, which in addition are under-equipped. Most of the bites occur in the rainy season. Almost 3 out of 4 victims are male and the average age of the patients is 28 years. These are mainly peasants bitten while working in the fields.

In 1993 a survey was carried out of households and health care centers in 6 states in Nigeria. The incidence was 155 snakebites per 100,000 inhabitants, close to 200,000 bites every year. In the states with a high incidence this lies between 250 and 400, whereas in the other states the incidence varies between 10 and 70 bites. The overall lethality is 9.8%. In more than half of the cases the victims consult traditional healers; in some of the states this proportion can reach and even surpass 90%. Overall one third of the patients consult a heath care center, where the preferred treatment is anti-venom serum, if it is available. A serum therapy is administered to 60% of the envenomed patients. While it is probably the most severe cases which benefit from this, it does not prevent a very high lethality of hospital patients with an average of 27% ranging from 4.8 to 41.6% in different states. In the Benue valley in the north east of Nigeria the annual incidence of snake bites can surpass 600 cases per 100,000 inhabitants. The morbidity lies above 150 envenomations per 100,000 inhabitants and the lethality is close to 15% in the absence of specific antivenom treatment; the bites are caused mainly by *Echis ocellatus* as in the whole of the north of Nigeria. The majority of the bitten people are young adult men engaged in agriculture or hunting. Inflammatory syndromes (pain and edema) are observed in 80% of the cases and local or systemic bleeding in 60% of the patients.

In the Northern Province of Cameroon, into which the Benue valley continues, a similar incidence of between 200 and 300 envenomations per 100,000 inhabitants has been observed. The lethality equally is close to 10% in the absence of antivenom treatment. *Echis ocellatus* represents 85% of identified snakes having caused bites. A standardized specific treatment has reduced the lethality to 1.5% of the bites. The population at risk consists of people between 15 and 44 year of age, predominantly male. Most of the bites occur during agricultural work. In the cotton growing area 40% of the bites take place during the 3 months of field preparation and cotton sowing.

In Gabon about one hundred envenomations are treated each year. The morbidity is low, but the lethality is at 2.5%.

In the Congo Republic the incidence lies between 120 and 450 snakebites per 100,000 inhabitants per year in the forest. In urban areas it is 11.5 bites per 100,000 inhabitants. Curiously the morbidity is 20 envenomations per 100,000 inhabitants and the lethality varies from 1 to 7%. Most of the patients consult traditional healers and it is likely that a considerable number of bite cases does not reach a hospital.

In rural areas of Kenya the incidence is 150 bites per 100,000 inhabitants per year and the morbidity lies between 5 and 70 envenomations. Barely 20% of the bites are symptomatic and 70% of the bites are treated with the help of traditional medicines. The lethality of hospital cases is 3.6% but can reach 9.5% in the absence of appropriate treatment. The mortality varies from region to region between 0.5 and 6.7 per 100,000 inhabitants per year. Less than half of the victims consult a modern health care center and 70% of those have first received a traditional treatment.

In Zimbabwe the morbidity is 3.5 envenomations per 100,000 inhabitants, which appears to be under-reported, with a lethality of hospital cases between 1.8 and 5%. As many men as women are treated and the mean age lies between 21 and 30 years.

In South Africa the overall annual morbidity is 35 per 100,000 inhabitants with a lethality of up to 5%. The mean mortality is estimated to be 0.5 per 100,000 inhabitants per year. Children below 10 years, mostly in the black population, represent 40% of the bites. In Johannesburg more than half of the envenomations treated in hospitals since the 1970s are induced bites. In the Natal Province the morbidity is estimated at 100 envenomations per 100,000 inhabitants per year. More than 90% of victims treated in health care centers have first been treated by a traditional healer.

The disorganization of health services in many countries because of economic crises or civil or military insecurity, makes it particularly difficult to render service to the patients. That is partly the reason why less than 30% of the patients are treated in a health care center.

In a population of 800 million people a million snakebites occur each year in Africa. The 500,000 resulting envenomations cause 20,000 deaths of which only half are known to the health services. In effect, the proportion of victims who consult a health care center after having been bitten by a snake, is estimated to be below 40%.

3.5.5 Asia

In Asia there is a great variety of situations because of the diversity of human activities and of snake species involved in accidents. The frequency of bites by elapids (*Naja, Bungarus*) is generally higher than in Africa. The viperids are represented by the Viperinae (*Daboia, Echis, Vipera*) as well as by Crotalinae (*Protobothrops, Deinagkistrodon, Calloselasma, Gloydius, Hypnale*).

In Japan generally the incidence of snakebites is very low, approximately 1 case per 100,000 inhabitants. However, the morbidity is much higher on the southern archipelagos of the country because of the tropical situation and the preponderantly agricultural economy. There close to 350 bites per 100,000 inhabitants

The Epidemiology of Envenomations

per year have been reported with a lethality of 0.7%. There are large variations between the islands and they are linked essentially to the cultivated surface and the agricultural techniques used. On Okinawa the morbidity lies at 35 envenomations per 100,000 inhabitants, 130 on Amamioshima and 580 on Tokunoshima. Close to 40% of the envenomations occur in the sugar cane plantations. The morbidity lies at about 120 envenomations per 100,000 inhabitants. In this country as in Taiwan, in Thailand and in the south of China and of Korea, the snakes causing most of the accidents are the *Trimeresurus* spp. and the *Protobothrops* spp., notably *P. flavoviridis*.

In Korea the lethality of hospitalized cases of the envenomations lies at 5%.

In Taiwan 15,000 to 20,000 bites are reported each year with close to 500 deaths. More than 60% of the bites occur during agricultural activities.

In continental China, where medical statistics are not accessible with the exception of rare medical publications, the lethality of hospitalized cases is 1%.

In Thailand about 10,000 envenomations are counted annually, which appears to be quite underestimated. *Calloselasma rhodostoma* is responsible for 35 to 40% of the bites, *Timeresurus albolabris* for 25 to 30%, *Daboia russelii* for 15%, *Naja atra* for 10%, *N. kaouthia* for 5 to 10 % and *Bungarus candidus* for 1%. *Trimeresusus macrops* could be implicated locally in a high percentage of bites, but generally it is considered as relatively rare. The sex ratio is even and the population at risk consists to 70% of active adults. More than 60% of the bites occur between noon and midnight with a peak at dusk. Half of the bites occur in or close to the houses. The lethality has been reduced by 90% with the use of antivenom serum and the standardization of the treatment protocol.

In Malaysia the incidence is 450 bites per 100,000 inhabitants per year and the morbidity lies between 130 and 300 envenomations per 100,000 inhabitants. The lethality of hospitalized cases lies below 0.5%. In half of the hospitalized envenomation cases the bites are caused by *Calloselasma rhodostoma*.

In the Philippines a very high mortality has been reported amongst rice farmers. For the whole of the population morbidity and mortality are unknown.

In Myanmar the incidence lies between 35 and 200 bites per 100,000 inhabitants with a certainly under-estimated morbidity of 35 envenomations per 100,000 inhabitants per year. The mortality is the highest in the world: 35 deaths per 100,000 inhabitants per year. *Daboioa russelii* is credited for 60 to 70% of the envenomations, the same as in certain eastern provinces of India. In these regions where less than 40% of the patients consult a health care center, the incidence lies between 200 and 500 bites per 100,000 inhabitants per year with a morbidity of 100 to 200 envenomations and a lethality of between 2 and 20 deaths per 100,000 inhabitants.

In Nepal more than 20,000 bites are treated each year by the medical services and more than 1,000 deaths are reported. More than half of the bites occur during the 4 to 5 months of monsoon. The accidents involve mainly active young persons. The lethality lies between 3.8 and 5% and the mortality is estimated at 5.6% deaths per 100,000 inhabitants per annum. Elapid bites represent 20% of the envenomations.

In India the incidence varies between 60 and 165 bites per 100,000 inhabitants per year. The morbidity reaches 70 envenomations per 100,000 inhabitants in the regions with a prevalently rural economy like in Maharashtra where the mortality is 2.4 per 100,000 inhabitants per year. The lethality of hospitalized cases lies between 1.5 and 20%. The bites of viperids occur two times more often than those of elapids. The former takes place in the afternoon and early evening between the month of June and November. The latter are distributed throughout the day/night cycle with an increase during the night, and throughout the year with a peak in June. Elapid bites occur more often in towns and in densely populated zones, while the viperids bite mainly in the rural regions during agricultural activities. The lethality is 25% for the envenomations by elapids and 6% for those by viperids.

In Pakistan the lethality of hospital cases is 0.5%. Envenomations occur in 65% of the bites and affect the lower member in 85% of the cases.

In Sri Lanka the incidence is 250 bites per 100,000 inhabitants per year and the mortality approaches 6 per 100,000 inhabitants with peaks in some places of 18 per 100,000 inhabitants. However, less than 20% of the patients consult hospitals for treatment. *Daboia russelii* and *Naja naja* are the most dangerous species with respectively 40% and 35% of deaths. *Echis carinatus* is present in Sri Lanka but appears to be rarely implicated in the accidents, while in India and Pakistan it is responsible for 80 to 95% of envenomations.

In Asia with a total population of around 3 billion inhabitants, the incidence of snakebites is estimated at 4 million accidents per year, half of which cause an envenomation. The number of deaths lies at 100,000 per year.

3.5.6 Australia and Oceania

In Australia around 3000 bites are counted each year. The incidence of snake bites lies between 5 and 20 per 100,000 inhabitants. The mean mortality is 0.03 per 100,000 inhabitants, four deaths per year. Most of the bites are inflicted by species of the genus *Pseudonaja*, which cause about half of the deaths, followed by *Notechis* spp. and *Oxyuranus* spp.

In Papua New Guinea the incidence varies between 200 and 525 bites per 100,000 inhabitants. The envenomations represent 1.1% of all the deaths and more than 20% of violent deaths, accidents and suicides.

The rest of the Pacific is free of venomous snakes with the exception of the marine snakes, which have a neurotoxic venom, but because of their habitat and their lack of aggressiveness they cause very few accidents.

With a population of 20 million inhabitants Oceania has 10,000 snake bites with 3,000 envenomations per year causing about 200 deaths.

3.6 The medical response to snakebites

The medical response to snakebites poses two types of problems: the speed of intervention and the efficacy of the treatment.

In the industrialized countries there are intensive care services as well as data banks on envenomations. They are regularly kept up to date at anti-poison centers or the equivalent in the respective country and give the practitioner access to information and advice.

In developing countries the mediocrity of the care offered and the distance between health care centers do not allow an early hospitalization of the victim, if at least his initial reaction is to go to a dispensary (Fig. 46). The mean delay of consultation is from 16 to 48 hours in Nigeria, depending on studies and districts, 12 hours in northern Cameroon and 7 hours in Brazil. In France it is largely below 2 hours.

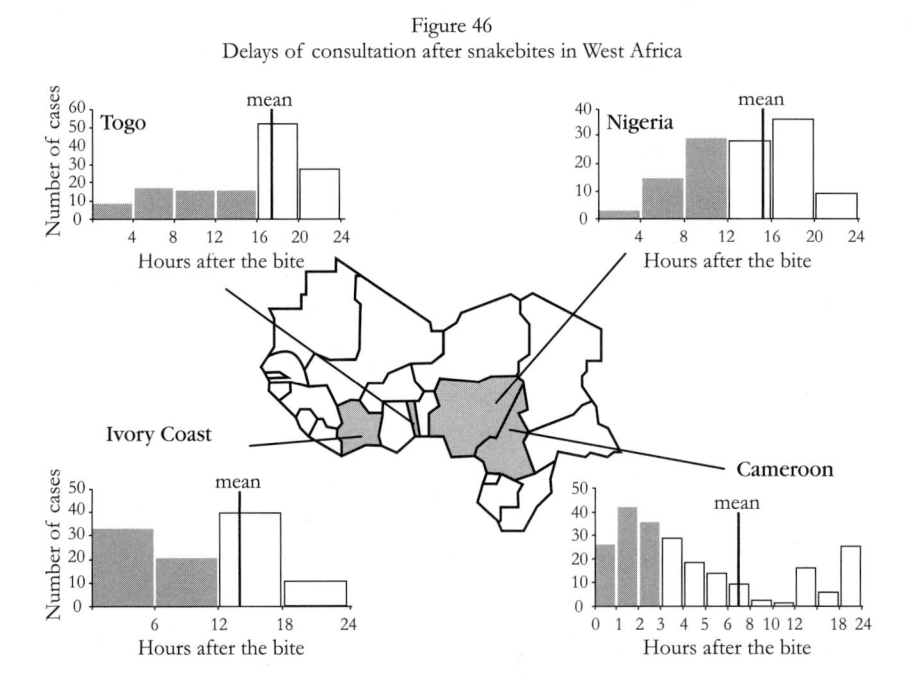

Figure 46
Delays of consultation after snakebites in West Africa

The accessibility of health care centers, the competition by the traditional healer, the cost of modern treatment (one to two times the average monthly income of a family in Africa) which is rarely covered by the state or the institutional sector, the available therapeutic resources of the health care center and the severity of the envenomation are the main criteria on which the decision of the victim or of his family and friends is based.

CHAPTER 6
Clinic and Treatment of Envenomations

Snakebites present a medico-surgical emergency and their frequency can in certain countries pose a real public health problem. This varies with the nature of the snake fauna and the state of the health care structures.

Also, the therapeutic approach to the envenomations has evolved over the last decades: the prognosis of survival is no longer the only one to be considered. Increasingly attention is given today to local and regional complications and to the duration of the treatment.

1. Symptomatology and Clinical Pathology of Envenomations

Neuro-muscular syndromes

In principle the injection of the venom is painless. There are exceptions like the bites by *Dendroaspis* spp. or certain *Naja* spp and also *Walterinnesia aegyptia*. Neuro-muscular symptoms appear early and rapidly develop into one of the syndromes that are characteristic of the envenomation by elapids (Fig. 47).

Often at the beginning of the envenomation the nonspecific clinical signs are taken as fear reactions and divert the diagnosis of the practitioner who has little experience with snakebites. The acute forms of stress can in effect be confounded with a beginning cobra envenomation and vice versa. However, the unobtrusiveness and the slow and progressive worsening of the symptoms should point at a diagnosis of envenomation rather than stress without envenomation. The symptoms of the latter generally are severe and complete from the start.

Figure 47
Chronology of an envenomation by elapids

0 → Varying pain

1/2 hour: Discrete local signs / Sensory and motor problems

1 hour: Dyspnoea

3 to 12 hours: Respiratory paralysis → Death

Local Inflammation

No clinical signs = No envenomation

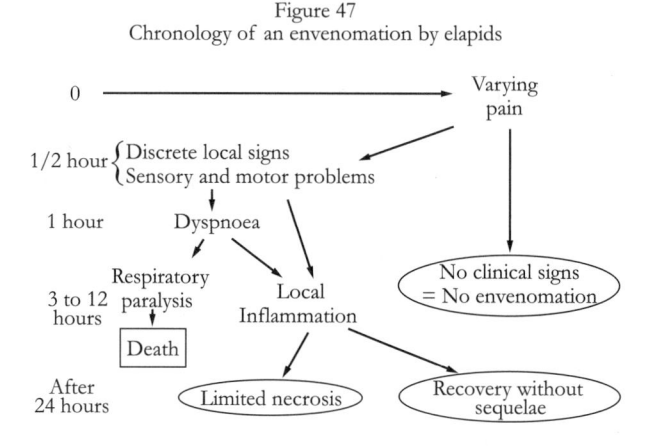

After 24 hours: Limited necrosis

Recovery without sequelae

1.1.1 Muscarine-like syndrome

The muscarine-like syndrome seen after the bite of *Dendroaspis* spp. (African elapids) is accompanied by objective signs that permit an early diagnosis. It is linked to the dendrotoxins or facilitating toxins, which enhance the release of acetyl-choline and thus the transmission of nerve impulses.

With the *Dendroaspis* spp. and certain *Naja* spp. the injection of the venom is paradoxically sometimes painful. Before the cobra syndrome described below develops, the patient will complain about diffuse sensorial problems. Generally this begins with diffuse "cold" sweating together with lacrimation and an abundant flow of saliva. Soon after a dysphagia appears, described as "a lump in the throat", which prevents the patient from swallowing his saliva. Nausea accompanied by epigastric pains and exceptional vomiting and diarrhea appear rapidly. Visual disturbances – spots, brief luminous stars or more prolonged "swimming before the eyes" – then the impression of double vision because of accommodation difficulties follows with auditory sensations, humming or whistling in the ears (tinnitus); these symptoms are not really severe, but cause a strong fear in the victim. The examination of the patient reveals a narrow constriction of the pupil or myosis and sometimes a trismus, in the form of a contraction of the jaws which is either spontaneous or is triggered by a stimulation of the palate with an instrument (e.g. tongue depressor).

1.1.2 The convulsive syndrome caused by the fasciculins

The fasciculins inhibit cholinesterase, which degrades acetyl-choline and interrupts the transmission of nerve signals. They act in synergy with the dendrotoxins provoking particular motor problems. These begin with superficial trembling, starting from the site of the bite and progressing along the member up to the body. This trembling gives way to shaking which follows the same path. Progressively they are transformed into cramps and finally spasms. The development towards convulsions or epileptic crises has never been reported in man and appears to be unlikely because of the weak concentration of the toxins.

1.1.3 The cobra or curare-like syndrome

The cobra syndrome is the logical continuation of the signs described above. The neurotoxins blocking the nerve transmission lead to a motor paralysis. The clinical picture is that of a more or less pure and rapid curarization depending on the specificity and the mode of action of the neurotoxin involved.

The cobra envenomation progresses rapidly. It may or may not be preceded by the symptoms described above, which can be observed in a blurred form in all elapid bites, and is accompanied by a progressive resolution of all the neuro-muscular signs.

After a procession of paraesthesias starting from the bite and progressing towards the body and the head, which are essentially sensorial (anaesthesia, stinging, tickling, feeling chilly) and hardly accessible to an objective examination, the first clearly visible physical symptom is the bilateral and symmetrical ptosis, the irrepressible closing of the eyelids. Almost simultaneously one observes the apparition of a trismus, the contraction of the jaws, either spontaneous or triggered by the stimulation of the palate by a tongue depressor, which is bitten forcefully by the patient. The sign of the captive tongue depressor is also observed in tetanus, which has a similar aetio-pathogensis. Slowly the patient loses altogether his ability to communicate. His voice becomes hoarse and then extinguished. Hypotension is obvious and sometimes develops into a state of shock. Dyspnoea appears; it is translated into the slowing of the respiratory rhythm and accompanied by a "thirst for air". The victim becomes somnolent and responds less and less to exterior stimuli. Cut off from his surrounds by not being able to communicate, the patient gives a comatose impression. However, he is conscious and understands all that is said around him without being able to react. He has more and more difficulties breathing and progressively asphyxiates until he dies from respiratory arrest. The progression towards the terminal stage can take 2 to 10 hours depending on the quantity of injected venom and the size of the victim.

This syndrome is not accompanied by any neuro-muscular or cerebral lesion. The terminal coma is a quiet coma during which the consciousness is never changed and which is the translation of the motor paralysis without sensorial effect.

The envenomation caused by snakes which have pre-synaptic neurotoxin of the type β-bungarotoxin is, in addition to the cobra syndrome described above, often accompanied by a myotoxicity, which leads to a muscle destruction. The biochemical examination shows a myoglobinuria and an increase of creatine kinases, intercellular enzymes, which are released by the destruction of the striated muscle fibers. Sometimes an acute renal insufficiency complicates the clinical picture.

Certain Australian elapids cause in addition severe haematological problems.

1.2 The viper syndrome

The viper syndrome, sometimes split into inflammatory and necrotic syndromes, is associated with pain, edema, skin problems and necrosis. Hematological problems are often present but constitute a distinct entity regarding its etiology and its course (Fig. 48).

The pain is immediate. It is caused by the penetration of the venom. It is always fierce, transfixing, sometimes causing fainting. It radiates rapidly towards hip or shoulder and precedes the other inflammatory symptoms. At first it is probably of mechanical origin (injection of the venom under pressure and into depth) but its persistence later is tied to the complex mechanisms of inflammation, notably the presence of bradykinin.

The edema appears less than half an hour after the bite. It is the first objective sign of the envenomation and as such needs to be followed with full attention. It is voluminous, hard and extended. It stretches along the bitten member and increases in

Clinic and Treatment of Envenomations

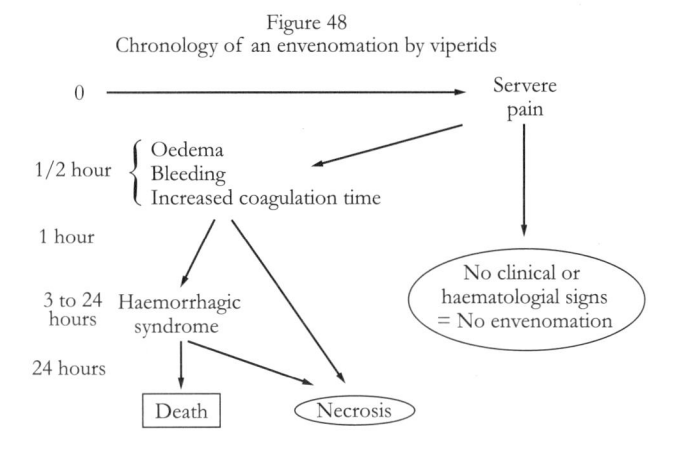

Figure 48
Chronology of an envenomation by viperids

```
0 ─────────────────────────────►  Severe pain
                                        │
1/2 hour   { Oedema                     │
           { Bleeding  ◄────────────────┤
           { Increased coagulation time │
                                        │
1 hour           ╱│╲                    ▼
                ╱ │ ╲           ┌─────────────────────┐
3 to 24  Haemorrhagic           │  No clinical or     │
hours    syndrome               │  haematological signs│
                ╲  │ ╲          │  = No envenomation  │
24 hours         ╲ │  ╲         └─────────────────────┘
              ┌──────┐  ╲
              │ Death│  (Necrosis)
              └──────┘
```

volume during the first few hours, stabilizing after 2 to 6 hours. A simple classification allows one to monitor its course and to modulate the treatment (Table XV).

The edema has a strong inertia: It decreases very slowly, which makes it poor indicator of clinical improvement and of healing. For a given species the extent of the edema is in proportion to the quantity of injected venom and therefore of the severity of the envenomation. Paradoxically the intra-compartmental pressure remains moderate even in the case of a spectacular increase of the volume of the edema. The muscle compression is limited to the muscle fascicles into which the venom was injected. The functional consequences generally are reasonably favorable. Measuring the intra-compartmental pressure allows one to evaluate the risk of a tissue anoxia through vascular compression (Volkmann syndrome) and to apply a surgical treatment, which, however, is very controversial.

The skin problems are linked essentially to the size of the edema and the existence of a hemorrhagic syndrome. The skin loses its elasticity, stretches and cracks, causing fissures, which generally are superficial but are sources of secondary infections and hemorrhages. The signs of hemorrhage (ecchymoses, petechiae, purpura, blood blisters) appear more slowly. The ecchymoses are a sign on which the severity can be judged.

Table XV
Clinical severity score

Level of severity (score)	Œdema	Bleeding
stage 0	Absent	Absent
stage 1	rises along the leg or the lower arm without reaching the knee or the elbow	bleeding at the point of the bite lasting more than one hour
stage 2	reaches the knee or the elbow	bleeding from other skin lesions than the point of the bite (scarification, wound)
stage 3	passes beyond elbow or knee without reaching shoulder or hip	bleeding from healthy mucosa
stage 4	reaches shoulder or hip	bleeding from untraumatized skin
stade 5	passes shoulder or hip	exteriorisation of an internal haemorrhage (haemoptyalism, haematemesis, melaena)

The necrosis is progressive. It first starts with a black dot that can become visible one hour after the bite. It extends both superficially and into the deep tissues. It continues as long as the venom is present in the body. Later in the absence of a secondary infection, which can evolve towards gangrene, the necrotic zone dries up and mummifies. The severity depends on the composition of the venom and the injected quantity. Some venoms, like that of *Bothrops brazili*, a crotalid of South American forests, have a strong proteolytic potential which causes a deep muscular necrosis without a sign of skin lysis. In all cases the muscle necrosis causes a considerable increase of the muscle creatine-phosphatases. The increase of these cytoplasmic enzymes above the pathological threshold is seen in 75% of patients with a complete or partial viper syndrome, viz. a large edema with or without necrosis. This rise of creatine-phosphatases is early and returns to normal within three days after the start of the regression of the edema.

Gangrene is a secondary complication of the tissue anoxia, and generally occurs after a tourniquet has been too tight and kept in place too long or after other local interventions. It does not show any notable clinical difference compared to the classical gas gangrene.

It is important to distinguish between a subcutaneous envenomation, which in its extension does not cause an intra-compartmental pressure, and an intramuscular envenomation that can be complicated by a Volkmann syndrome causing tissue anoxia and a high risk of gangrene.

1.3 Necrosis through cytolysis or the cytolytic syndrome

The differential diagnosis between the viper syndrome and the necrosis caused by the venom of certain elapids (notably *Naja nigricollis* and *N. mossambica*) can be difficult. The inflammatory syndrome caused by the venoms of the elapids is generally absent or indistinct. The phospholipases A_2 that abound in elapid venoms, have different substrates and a more neuro-muscular tropism. They do not have proteolytic enzymes affecting tissue structures or certain metabolic systems like coagulation or the synthesis of bradykinin and thus cause considerably less symptoms of inflammation, particularly pain and edema. The elapid venoms generally cause necroses with a slower development and a more limited extent than those of the viperids. The necrosis is dry. It mummifies rapidly and the necrotic lump falls off spontaneously if it has not been removed surgically during the stabilization of the wound.

1.4 The hemorrhagic syndromes and haemostasis problems

The bites by vipers or crotalids appear clinically in the form of a hemorrhagic syndrome, either primary (in rarer cases) or secondarily, following a patent thrombosis syndrome or one that was not diagnosed because of the instability of the thrombus. The etiological diagnosis of the hemorrhagic syndrome is never easy, even with the help of a specialized laboratory. However, in most cases it is a rapid afibrinogenaemia essentially due to the trombin-like enzymes or the prothombin activators.

1.4.1 Clinical diagnosis

The primary thrombosis syndrome is rare, except with certain species (*Bothrops lanceolatus* and perhaps *B. brazili*). The venom of these crotalids, respectively from Martinique and Amazonia, causes a sudden coagulation of the blood inside the vessels. Even though the thrombi formed in this way are abnormal and instable, they are sufficient to cause disseminated embolisms and infarction in the body which can affect any organ, like the heart (myocardial infarct), the kidneys, the lungs or other vital organs. Death or severe sequelae can be observed. In addition to heart and cerebral lesions with a serious prognosis, kidney problems occur most frequently. The visceral complications and their possible sequelae are described further below.

The hemorrhagic syndrome most often establishes itself insidiously. Generally it begins with a discrete and permanent flow of blood from perforations caused by the poison fangs. The victim shows this bleeding at his arrival, which *a priori* does not alarm the uninformed practitioner. It is important to take into account the delay between the bite and the arrival at the health care center for measuring the length of time of this bleeding. The persistence of this sign demonstrates a blood hypo-coagulability at least locally. At this stage the haemorrhagins of the venom alone can be responsible for the flow of blood, which may seem insignificant. This situation may remain stable for several hours and even several days.

Bleeding can occur away from the bite, in a recent wound caused by an intentionally therapeutic intervention like scarifications, incisions or the relieving of a voluminous edema. The wound can also be spontaneous as a result of a severe increase of the volume of the integuments, which become stretched and fissured.

Ecchymoses are subcutaneous suffusions of blood. They appear within 24 or 48 hours and are the first sign of exteriorization of the systemic hemorrhagic syndrome. They are a sign of the severity of the envenomation and have a serious prognostic with a sensitivity of 85% and a specificity of 100%.

At a more advanced stage hemorrhages can appear in old scars of reputedly healed wounds.

Clinic and Treatment of Envenomations

Finally the hemorrhages appear on the mucosae and the healthy skin in not previously injured places. The failure of coagulation provokes the escape of blood from the vessels – or extravasation - which appears as a purpura (eruption due to intradermal hemorrhages), unstoppable bleeding from nose and mouth, blood in sputum, urine and stools, and even cerebral and deep visceral haemrrhage. The development towards a severe anaemia or a hypovolaemic shock can bring about the death of the patient within a few days.

1.4.2 Clinical pathology – additional tests

Often there is a disagreement between the clinical pathology and the clinical picture. The appearance of the clinical signs can be retarded considerably in relation to the biochemical disturbances. Biochemically the onset of the hemorrhagic problems is sudden, within minutes or hours of the bite; the fibrinogen is consumed early and a fibrinolysis can rapidly complicate the situation. In certain cases, bites by *Echis* spp. for example, the exteriorization of the hemorrhagic syndrome may appear only several days after the bite. Often a persisting bleeding from the bite wound or from scarifications applied after the accident, are the only clinical manifestation of the hemorrhagic syndrome.

Whereever this is possible, a complete range of hematological tests, including a haemogramme for evaluating the severity of the anaemia as well as coagulation tests, should be carried out before and after each therapeutic intervention: antivenin, symptomatic treatment. Most of the envenomation accidents occur far from health care centers which have a laboratory; some simple clinical pathology tests carried out at the bedside of the patient, can be very useful to guide the treatment and monitor the patient.

The coagulation time in an untreated tube confirms the hemorrhagic syndrome and allows to judge the quality of the clot that forms. This test is simple, quick and very reliable (Fig. 49).

Five milliliters of venous blood are taken with a syringe and emptied into an untreated tube, without anticoagulant and preferably new, never having been in contact with soap or a detergent, or carefully washed, rinsed and dried. The tube is placed on a stable stand for 30 minutes. After this period of incubation the contents of the tube are examined: if it has not clotted or if the clot is incomplete,

Figure 49
Coagulation test in an untreated tube

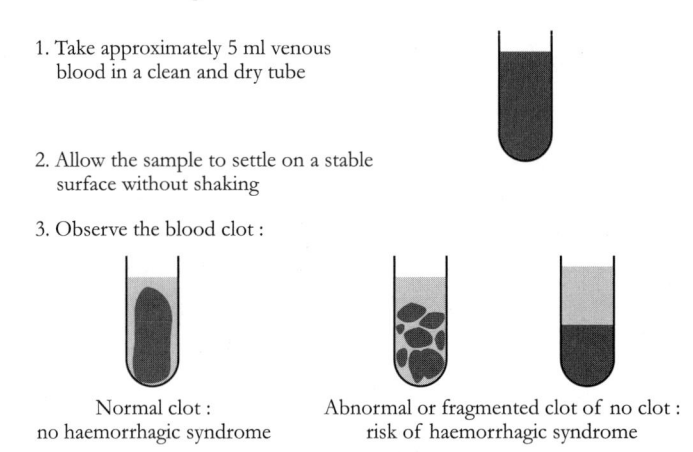

1. Take approximately 5 ml venous blood in a clean and dry tube

2. Allow the sample to settle on a stable surface without shaking

3. Observe the blood clot :

Normal clot :
no haemorrhagic syndrome

Abnormal or fragmented clot of no clot :
risk of haemorrhagic syndrome

friable or quickly dissolved after a light shaking of the tube, there is a coagulation problem. This test can be repeated as long as the patient is under observation; it will provide information on the development of the hemorrhagic syndrome.

The other tests, if they are available, allow a more precise diagnosis of the mechanisms of the coagulopathy. The fibrinogen level has collapsed in all the cases. The prothrombin time and the cephalin time provide information on the clot and its state. Measuring the fibrin degradation products (FDP) allows the detection of a fibrinolysis. The precise identification of the fragments allows one to distinguish between a reactional fibrinolysis (FDP_n), which results from the destruction of stabilized fibrin of a normal clot, or a primary fibrinolysis or the degradation of an abnormal clot consisting of soluble fibrin caused by the action of the venom (FDP_g). Several tests are used to evaluate the intensity of the fibrinolysis (time of lysis of a clot of diluted blood, von Kaulla's test, or the time of lysis of a clot of euglobulins, measuring the protease activity of plasma, Astrup's platelet test, etc.). Sometimes they are difficult to interpret in the particular context of the presence of numerous and varied proteolytic enzymes in the blood of the patient. The platelet count is down in the case of a disseminated intravascular coagulopathy (DIC syndrome), which is diagnosed frequently in viperine envenomations caused by species that have no thrombin-like enzymes in their venom. It is rare in other cases. The bites of *Echis ocellatus*, which has a prothrombin-like enzyme in its venom, causes a DIC syndrome in 62% of the patients, while an isolated fibrinolysis is found in 20% and an afibrinogenaemia in 15% of them.

Measuring the intra-compartmental pressure allows one to follow the progression of the edema and to determine the need for an eventual surgical intervention.

Ultrasonography equally is very useful for the surveillance of the edema as for the adaptation of the treatment. An increase of the echo from the muscles points at a vascular failure and a high intra-compartmental pressure.

The thrombo-elastogramme is sensitive and detects early disturbances (Fig. 50). They indicate the risk of a hemorrhagic syndrome without though formally identifying its etiology. The length of the segments r and k on the thrombo-elastogramme inform on the thromboplastin formation, that is on the delay before the formation of the clot, and on its quality. The amplitude of the two arms of the fork (amx) provides information on the structure of the clot and its stability. In the case of a beginning viperine envenomation, r and k are much shortened and amx increased. Thereafter, as the haemostasis factors have been used up and because of the inability of the blood to coagulate, amx is reduced and disappears completely. However, the thrombo-elastogramme does not allow an etiological diagnosis of the observed hemorrhagic syndrome; it is mainly used for showing its existence and for monitoring it.

The anemia develops quickly. The patients presenting a hemorrhagic syndrome lose 25% of their hemoglobin in a few hours. It is a hypovolemic anemia, which is proportional to the duration and severity of the hemorrhagic syndrome

The number of leucocytes increases in most of the envenomations, except for the envenomations by the Australian elapids, in which the leucocytes diminish. The increase is in the order of 20 to 30% and affects mainly the polynuclear neutrophils and eosinophils. They normalize rapidly after the patient has recovered.

Systematic and repeated tests are necessary for monitoring kidney function at least during the first days for proteinuria, haematuria, uraemia and creatininaemia, as

Figure 50
Diagnosis of haemorrhage problems by thrombo-elastogramme

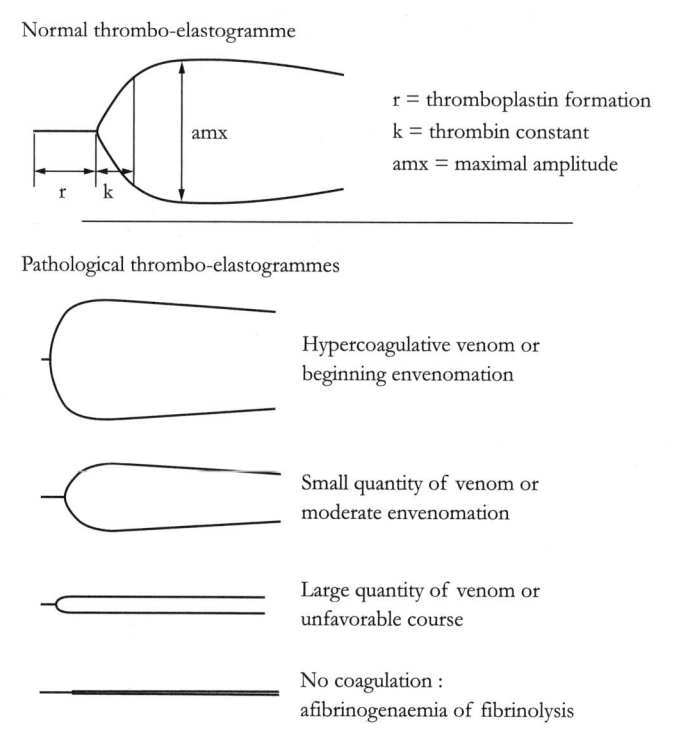

Normal thrombo-elastogramme

r = thromboplastin formation
k = thrombin constant
amx = maximal amplitude

Pathological thrombo-elastogrammes

Hypercoagulative venom or beginning envenomation

Small quantity of venom or moderate envenomation

Large quantity of venom or unfavorable course

No coagulation : afibrinogenaemia of fibrinolysis

---

well as the blood N-acetyl-β-D-glucoseaminase, which is a cytoplasmic enzyme that is increased by the destruction of the cells of the kidney tissues. Levels of this enzyme can increase well before an acute renal insufficiency becomes apparent.

Testing for blood in the urine, simply with the help of a strip, for detecting a beginning haematuria, measuring the haematocrit in a heparinized capillary tube to evaluate the dilution of the blood, and determining the coagulation time in an untreated tube, can be of great help in isolated health care centers.

1.5 Haemolysis

In general the haemolysis syndrome remains subclinical. The anemia remains discrete, it is exceptional to see an icterus and recovery is spontaneous. However, it is not impossible for certain visceral, particularly renal, complications to be caused by the haemolysis. Other factors can also be involved in such developments, which occur several days or weeks after the accident.

1.6 Visceral complications

1.6.1 Complications of envenomations by elapids

Envenomations by African elapids do not change any other function than respiration. No neurological, cardiovascular or renal sequelae have ever been described after correctly treated envenomations.

The spray of venom of the spitting cobra into the eyes causes a severe conjunctivitis generally without sequelae. However, if the treatment is too late or too aggressive, it can lead to permanent corneal lesions with reduced visual acuity or even blindness.

Most often therefore the complications are iatrogenic or nosocomial.

1.6.2 Complications of envenomations by viperids

Sequelae occur in 1 to 10% of envenomations by viperids. They are linked to necrosis, which in the end can necessitate an amputation, or to the thrombosis syndrome that can lead to visceral infarction away from the site of the bite. This is seen frequently in envenomations by *Bothrops lanceolatus* in Martinique, in which 30% thromboses are observed against 3% hemorrhagic syndromes. In addition this species causes 5% necroses.

Two etiologies can explain the nephrotoxicity of viperine envenomations. Clinically they are different and this allows the treatment to be adjusted.

- *A direct nephrotoxicity* is observed in 25% of the bites by certain viperids, notably by *Bothrops moojeni* or *Daboia russelii*, immediately after the bite. It can appear in the absence of a patent systemic envenomation. The functional disturbances are retarded, sometimes remain indistinct and only the reduced glomerular filtration detected by the appearance of N-acetyl-β-D-glucoseaminidase in the urine allows one to establish the diagnosis within hours following the bite. The other biochemical parameters (protein, urea, creatinine) remain normal for a long time. Histologically the epithelium of the proximal tubuli is destroyed in its superficial part. The lesions are similar to those seen in the nephrosis syndrome. They affect the epithelial cells of the nephron, the functional unit responsible for the filtration of the urine, and the structural fibers that hold it in place, but never the basement membrane. These lesions probably are caused by the enzymatic activity of the proteases and the phosphlipases A_2 of the venom, which affect the epithelial cells of the proximal tubuli. The immunotherapy allows to achieve a recovery in 3 to 5 days with complete tissue repair. If in contrast an acute renal insufficiency occurs, it causes 5% of the deaths due to *D. russelii*.

- *A non-specific nephrotoxicity* is observed with the venoms of many viperids. The symptoms appear slowly and progressively but always are clinically manifest. The biochemical signs are in contrast early and many and are easily detected, even in developing countries (proteinuria, uraemia, creatininuria, haematuria). The histological lesions are distributed randomly along the tubulus and the glomerulus leading to their complete destruction. The basement membrane is always affected. This mechanism is totally independent of an immune pathological reaction, shown by the absence of deposits of immunoglobuline or complement on the glomerulus. Most likely it is linked to the hemorrhagic syndrome, notably disseminated intravascular coagulation, and to the haemolysis, which provoke a persistent ischemia of the blood capillaries of the renal parenchyma and a deposition of fibrin on the vascular endothelium. A vascular spasm has sometimes been incriminated in cases of envenomations by *Bothrops* spp.

This type of complication is also encountered following bites by *Atractaspis* spp. and *Bitis gabonica*. In contrast, renal complications are to our knowledge not observed after an envenomation by *Echis* spp. except for those that can accompany a hemorrhagic or allergic syndrome.

The nephrotoxicity of an immune-allergic reaction, notably of iatrogenic origin, is not excluded, but it is exceptional and appears to be unlikely with the presently produced highly purified antivenins.

1.6.3 Cardio-vascular complications

The venom of the *Atractaspis* spp., snakes in Africa and the Near East, is strongly cardiotoxic due to the sarafotoxins. These peptides cause an increase in the force of the contractions of the heart (positive inotropism) and cause a vasoconstriction with an auriculo-ventricular block. Clinically the envenomation is characterized by a transitory hypertension with an increased diastolic pressure and characteristic disturbances of the electro-cardiogramme: a prolonging of the PQ segment by more than 0.25 second and a flattened T wave which later becomes inverted in certain divergences. The bradycardia is regular and regresses in three days except that complications can cause the death of the patient.

The taicatoxin of the venom of certain Australian elapids (*Oxyuranus* spp.) causes a sinusoid bradycardia with atrio-ventricular block and an inversion of the T-wave in one third of the patients.

1.7 Eye injuries caused by spitting snakes

Spraying of venom into the eyes occurs often. It is done by the African spitting snakes (*Naja nigricollis, N. katiensis, N. mossambica, N. pallida, N. crawshayi* and *Hemachatus hemachatus*) and certain populations of Asian cobras (*N. sputatrix*). The venom is projected forcefully in the form of small droplets. It can reach a target situated at two meters distance. When the eyes are hit, there is immediately a strong burning sensation like in conjunctivitis. A symptomatic local analgesic, anti-inflammatory and antiseptic treatment is sufficient if it is applied in time and after a thorough rinsing of the eyes with water, physiological saline solution (salt in water 9 g·l^{-1}) or artificial tears.

1.8 Identification and titration of the venom

The titration of venom circulating in the body of the snakebite victim has for a long time been the object of research (Table XVI). It is not so much the specific

Table XVI
Principles and effectiveness of techniques used for the detection and titration of venoms dans in living tissues

Date	Name and principle	Sensitivity ($\mu g \cdot l^{-1}$)	Delay of response	Remarks
1950	Biological test: measuring the LD_{50} from blood or tissues of the victim	500	24 h.	Does not allow identification of the venom
1965	Immunodiffusion: precipitation of the antigen/antibody complex in a gel	100	24 h.	Simple technique which does not require complicated apparatus or reagents
1970	Agglutination: agglutination of erythrocytres or latex particles to which the venom has been fixed	50	2 h.	Reagents needing prior preparation
1975	Radio-immuno-essay: immunological reaction detected by radio isotopes	< 1	20 h.	Costly and complex reagents and apparatus
1975	ELISA: immuno-enzymatic method	2	2 h.	Simple and relatively inexpensive technique; needs reagents prepared in advance
1980	EIA: immuno-enzymatic method revealing the immune complex by a colour reaction	0,5	1 h.	Higher sensitivity than the preceding method because of a colour reaction
1990	Mass spectrometry: measure of the molecular mass after Ionisation et volatilization of the sample in a vacuum	< 10^{-6}	2 h.	Technique cannot be used for titration ; requires specific calibration marked by an isotope

diagnosis of the species incriminated in the bite, but rather the precise determination of the quantity of venom in circulation and its diminution under the treatment. It allows monitoring the development of envenomation and treatment.

The first methods lacked sensitivity and precision. The blood of the victim, a homogenate of tissues from the site of the bite or a saline solution in which organs of the victim had been incubated to extract the venom from them, were injected into a mouse to evaluate the toxicity of the extract; depending on the symptoms shown by the mouse one could deduce the nature of the toxin.

Immunological methods have replaced these rudimentary techniques. They consist in using the affinity of an antibody for an antigen from which it has derived. This specificity and the sensitivity of the reaction were first exploited in precipitation reactions (Ouchterlony tests and derivatives). The samples, blood or extract of tissues from the victim (containing the antigens), are placed in a gel close to the antibodies. While diffusing through the gel the two substances meet each other and form an insoluble complex that precipitates in the gel and becomes visible to the naked eye. This precipitation reaction is marked by an opaque arch in the translucent gel and allows one to confirm the presence of the venom and to identify its origin.

This property has been improved by various methods. The fixation of the antibodies onto particles like red blood cells or microscopic latex beads allows the demonstration of the phenomenon without a gel. Using radio isotopes for marking the antibodies allows the detection of invisible antigen-antibody complexes. The immuno-enzyme techniques (enzyme-linked immunosorbent assay, ELISA, or enzyme immuno assay, EIA) use the specificity of an enzymatic reaction to improve the detection of the reaction by measuring the quantity of liberated substrate and the production of a particular colour (cf. Fig. 31). These different reactions are sufficiently sensitive to be proportional to the concentration of the venom and this allows a specific and precise measurement. Mass spectrometry could improve the sensitivity but it remains a technique that is poorly adapted for diagnosis.

Today the immuno-enzyme methods are used to evaluate the elimination of the venom during the course of the treatment. It is now possible to follow the curve of elimination of the venom and to adapt the immunotherapy according to the results.

2. The Treatment of Envenomations

The severity of an envenomation is caused by the action of the toxic components, which has the following chronology:

- *The postsynaptic neurotoxins* act first and cause a respiratory arrest after a delay of between 30 minutes and 8 hours after the bite.

- *The pre-synaptic neurotoxins* are slower but produce similar symptoms leading to a respiratory paralysis in 1 to 12 hours.

- *The hypovolaemic shock* caused by the venoms of viperids is very early and occurs between 30 minutes and 2 hours.

- *The cardiotoxic effect* of certain venoms of Australian elapids or of *Atractaspis* spp. occurs in 2 to 12 hours.

- *The hemorrhagic syndrome,* with exceptions, is delayed by several hours or days; death occurs through anaemia or hypovolaemic shock after 8 hours to 6 days after the bite.

- *Necrosis or gangrene* occurs much later and is observed between 3 days and 1 month after the bite.

- *Visceral complications* appear 15 days to 2 months after the bite except when caused by the bite of *Bothrops lanceolatus*, in which case they appear within a few hours.

The intravenous injection of venom is rare but has particularly severe repercussions because of the rapid onset of the symptoms (30 minutes to 1 hour) and of their severity (cardiotoxicity of the cytotoxins and of the phospholipases).

2.1 First aid

These are immediate interventions at the place where the bite took place, and these should be recommended to peripheral dispensaries or those of commercial enterprises as well as to poorly equipped health care centers (Table XVII). Above all it is important to avoid all aggressive treatments that could burden the vital or functional prognosis. Tourniquet, incisions, cauterization, which have not proved their advantages or their efficacy, sometimes can cause dreadful complications.

Table XVII
First aid to be rendered after a snakebite

Calm the victim and reassure the accompanying persons
Disinfect the wound
Immobilize the bitten limb
Additional medical treatment :
- pain = analgesic
- œdema = anti-inflammatory
- bleeding = haemostasis
- neuro-muscular problems = antihistamines
- respiratory problems = artificial respiration

Evacuate the patient to a health care center

Without delay the wound has to be cleaned carefully and the evacuation of the patient to be organized.

While a tourniquet should not be placed because of the toxic and septic complications, it has been demonstrated experimentally and also clinically that the interruption of the lymphatic circulation draining the bitten limb, can reduce significantly the diffusion of the venom (Fig. 51). For a long time this practice was reserved for the neurotoxic envenomations caused by the elapids for which this protocol appears to have proved itself. It has been argued that in the case of bites by viperids the edema rapidly increasing under the bandage could hinder the blood circulation and cause an anoxia as if a tourniquet had been applied. The practice appears to weaken these fears (Fig. 51) and as long as it is carefully monitored, it is now stipulated to apply a bandage in all cases of snakebite. The pressure of the crepe bandage has to be so much weaker as the edema is more voluminous. Consequently many specialists recommend placing a tight crepe bandage and to immobilize the limb during the whole period of evacuation to a health care center.

Figure 51
Effect of a pressure dressing on the diffusion of the venom

Experimental elapid envenomation (Sutherland et al., 1979)

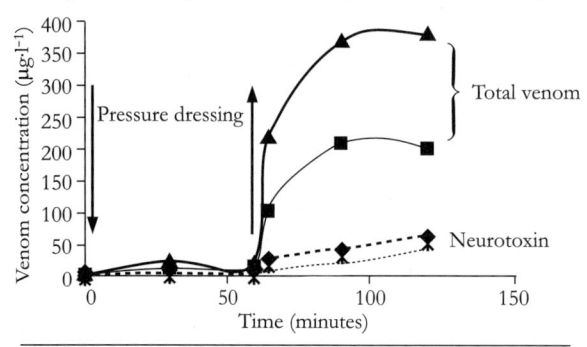

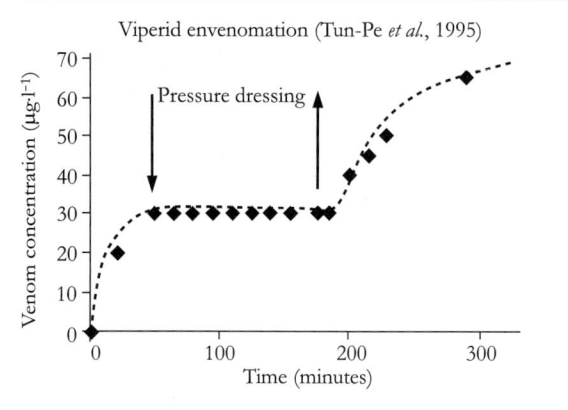

Viperid envenomation (Tun-Pe et al., 1995)

However, the release of the venom can be as sudden as after loosening a tourniquet. This makes it necessary to have a specific treatment ready to be used at the moment of taking off the bandage to neutralize the venom that is suddenly released into the body.

In the presence of neurotoxic signs (paraesthesia, fibrillation) the injection of corticoids and if the local signs are severe, the administration of an analgesic in combination with an anti-inflammatory, can be considered, as long as this does not delay the evacuation any further.

At this point in the logistics the availability of an antivenin or even more of a life support unit, cannot be expected. Their availability at this level can be envisaged in industrialized countries only.

2.2 Admission and additional treatment

The admission has the double objective of re-assuring the victim and of establishing as quickly as possible his actual state (Fig. 52).

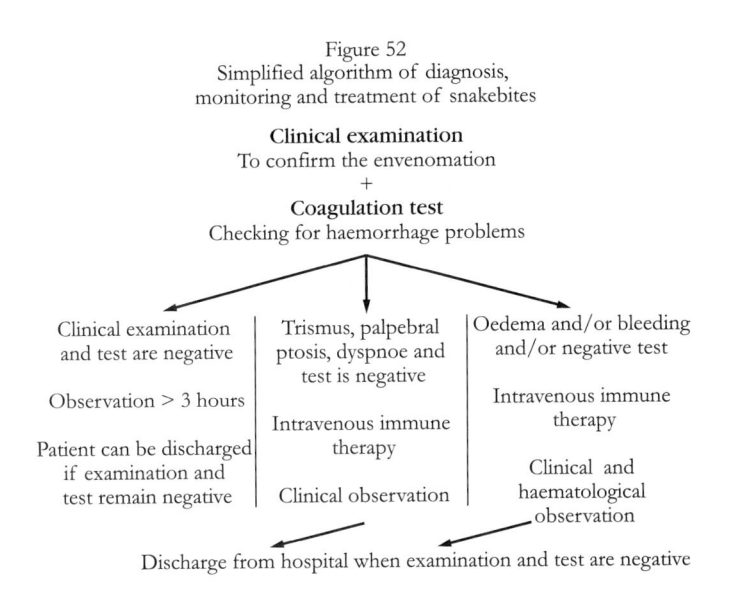

Clinic and Treatment of Envenomations

Figure 52
Simplified algorithm of diagnosis, monitoring and treatment of snakebites

Clinical examination
To confirm the envenomation
+
Coagulation test
Checking for haemorrhage problems

Clinical examination and test are negative	Trismus, palpebral ptosis, dyspnoe and test is negative	Oedema and/or bleeding and/or negative test
Observation > 3 hours		Intravenous immune therapy
Patient can be discharged if examination and test remain negative	Intravenous immune therapy	Clinical and haematological observation
	Clinical observation	

Discharge from hospital when examination and test are negative

In principle the clinical examination allows one to distinguish immediately between a viperine envenomation with its striking symptomatology – pain, inflammation, bleeding or even necrosis – from a cobra envenomation that is dominated by neuro-muscular problems. This distinction is theoretical, as one can sometimes observe inflammation and necrosis after bites by elapids and a paralysis in a viperine envenomation. However, these paradoxical syndromes are more limited and their clinical expression is very much different and will be described further below.

The interrogation evaluates the delay between the bite and the arrival at the health care center and identifies the treatment and therapeutic measures already applied.

The coagulation test in an untreated tube, already described above (cf. Fig. 49), is a useful precaution in countries where viperids are known to occur and it has no contra-indication. It is equally important for the diagnosis and for the further monitoring of the course of the envenomation in cases of the hemorrhagic syndrome.

Placing an intra-venous catheter should be a matter of routine.

The administration of a type H^1 antihistamine (chlorpheniramine: Polaramin®) can be justified on one hand because of its sedative effect that re-assures the victim, and on the other hand because of the antitoxic and potentiating effects, which although not explained have been confirmed in many experiments and clinical observations.

2.3 Immunotherapy

This remains the treatment of choice. It is the only etiological therapy and should be administered as soon as possible in sufficient dose and by the intra-venous route. The principle has been discussed in a preceding chapter where it was demonstrated that the immunotherapy should be considered as a comprehensive treatment of the envenomation and not only as an antidote to the lethal effects of the venom, which often are aleatory.

The indication of an immunotherapy should be considered in all cases of symptomatic envenomations irrespective of their supposed severity ((Table XVIII).

In effect the secondary effects of the allergic type, which have been put forward by some practitioners, have become exceptional since the advent of the purified

Table XVIII
Indications for the immune therapy

1. **Venomous snake identified with certitude**

2. **Envenomation confirmed clinically :**
   - intense pain
   - extensive oedema
   - cardio-vascular shock or falling arterial pressure
   - respiratory problems
   - neuro-muscular problems (ptosis, shaking, contractions, paralysis)
   - persistant local bleeding or spontaneous haemorrhages

3. **Positive coagulation test**

antivenin immunoglobulin fractions. The secondary effects are in any case treated more easily than severe envenomations. Also the development of an envenomation often is unpredictable. Finally it has been shown that even in benign envenomations the immunotherapy reduces significantly the time of hospitalization.

The earlier it is given the more effective is the immunotherapy. However, a long delay between the bite and the beginning of the treatment should not lead to its exclusion, as it is not possible to set a time limit beyond which the immunotherapy is no longer active. The posology will in any case take into account the delay before the start of the treatment.

The intravenous route is evidently the most appropriate one as it is the quickest one and leads to a maximum of efficacy.

The slow perfusion of antivenin diluted in glucose or salt solution has its partisans: the diffusion of the antibodies is constant and prolonged. Also it can be interrupted in the case of adverse reactions. The direct injection allows one to quickly obtain a high concentration of antibodies. The benefits of the one or the other technique have never been studied precisely. One can think that in the presence of a severe envenomation the direct intravenous injection would have a quicker and perhaps more decisive effect. The risk of side effects is not increased by this treatment: which ever the mode of administration, severe side effects remain below 0.5% while benign side effects are in the order of 5%. In the one case as in the other the antigens of the venom are attracted towards the blood compartment where the antibodies are. They will bind together to form immune complexes, which will be destroyed by the body in the organs of the immune system. In developing countries the direct injection has the further advantage of lowering the cost of the treatment by reducing the injection materials.

At present the recommendation is to administer the quantity of antibodies sufficient to neutralize the number of LD_{50}, which are on average in the venom glands of the snake in question, 5 ml in the case of *Vipera aspis*, 10 ml for a bite of *V. ammodytes* and 20 ml in most of the other cases. This dose will have to be repeated depending on the development: two hours later in the case of an aggravation or in the absence of an improvement, six hours after the first administration in the case of disquieting

Figure 53
Protocol for the immune therapy

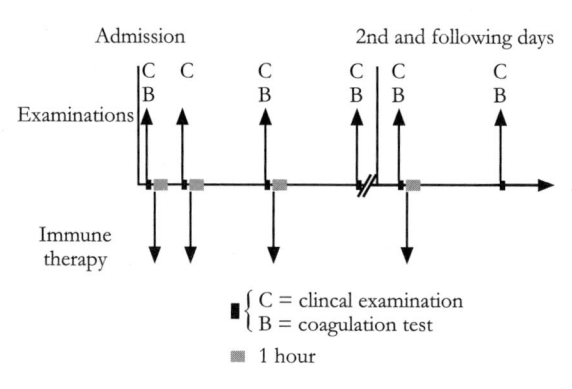

clinical or biochemical signs, neuro-muscular or hemorrhagic. The repetition every six to eight hours on the first day and later depending on the development during the following days is decided by the medical team according to their experience and the state of the patient (Fig. 53).

2.4 Medical treatment

2.4.1 Treatment of the cobra and muscarine-like syndromes

The beginning of a dyspnoea, presumed to be due to a respiratory paralysis (severe cobra envenomation), neccessitates the use of artificial respiration, eventually with endotracheal intubation. This needs to be continued as long as spontaneous respiration has not returned, which may take several days and even several weeks. A tracheotomy should be avoided if at all possible: It is rarely necessary and often develops into mechanical or infectious complications which can be serious and debilitating, particularly in developing countries.

Some authors administer neostigmine, which appears to potentiate the effect of the antivenin. Prostigmine® can be given in tablet form or as injection in the dose of 80 mg·kg^{-1}. Neostigmine and above all edrophonium (133 mg·kg^{-1}), which is a short-acting anticholesterase drug, are even more effective if their dose is fractionated. The slow intravenous injection of atropine (0.6 mg for an adult or 50 µg·kg^{-1} for a child) followed by edrophonium (Tensilon®: 10 mg for an adult or 0.25 mg·kg^{-1} for a child) allows one to discover the effect of a cholinesterase present in the venom. The treatment will consist in the administration of a long-acting anti-cholinesterase like neostigmine.

Atropine has been shown experimentally to be very effective against the mamba venom. The action of atropine at the dose of 8 $\mu g \cdot kg^{-1}$ is slow but long lasting but there is quite a risk of overdosing. In the case of a light envenomation the drug can be administered *per os*, otherwise by the parenteral route: 10 mg and thereafter 1 mg every 5 min until the symptoms have disappeared. Scopolamine can also be used as well as its synthetic derivatives, which have not proved to be of greater efficacy or to be easier to use. Atropine potentiates the curare-like effect of the neurotoxins and it is risky to use it in cobra envenomations.

The drug 3,4-diaminopyridine enhances the release of acetylcholine from the presynaptic membrane by blocking the potassium channels and leaving open the calcium channels. Experimentally used at a dose of 2 $\mu g \cdot kg^{-1}$ intravenously, it has been used successfully for treating the neurotoxicity of certain β-neurotoxins, notably that of *Bungarus fasciatus*.

Trypsin has been recommended as local injection with the aim of destroying the venom. Experimentally it appears to be active against the venoms of the elapids, provided it is administered in less than 10 minutes after the penetration of the venom.

Chlorpromazine (Largactil®) helps to obtain a powerful sedation.

2.4.2 Treatment of the haemostasis problems

The appearance of a hemorrhagic syndrome during an envenomation poses a poor prognosis.

2.4.2.1 Substitution therapy

Transfusions (full blood, plasma, PPSB, fibrinogen) only give an illusory solution as long as the venom is present in the body, as the supply of new substrate only re-activates the enzymatic action of the venom. Consequently the exclusive use of a transfusion cannot be justified, except as last resort to compensate for extreme losses when they constitute a major risk factor. A blood transfusion can also be envisaged in the case of a massive haemolysis which is the exception, even in the cases of an envenomation by elapids which have venoms rich in phospholipases. Such a therapy can in any case not be viewed in isolation.

The injection of vitamin K is of no interest in an emergency treatment. In the longer term it enhances the restoration of the factors of the prothrombin complex and in that capacity its therapeutic use is justified during the convalescence period.

2.4.2.2 Chemical inhibition of the venom

The enzyme inhibitors sometimes are an efficient resource, but they have to be used with many precautions.

Heparin had a great reputation, particularly in France, based on isolated and probably poorly evaluated experiments and clinical successes. Heparin binds to antithrombin III and exerts an inhibiting action on the serine-proteases that participate in the coagulation. It intervenes therefore in all the phases of coagulation: thrombin formation and fibrin formation. In addition it allows the release of the tissue plasminogen activator and thereby enhances the fibrinolysis. At a low dose heparin inhibits mainly the Stuart factor. The inhibition of thrombin occurs only at doses that are 5 to 10 times higher. Also heparins with a low molecular weight have practically no inhibiting effect on thrombin. The thrombin inhibiting action needed for the treatment of the disseminated intravascular coagulation (DIC) syndrome therefore requires relatively high doses of heparin of high molecular weight. The hemorrhagic syndrome observed in snake envenomations is generally regarded as a DIC syndrome and for it a heparin therapy appears to be justified.

In practice it has been shown that the inhibiting power of heparin on most of the thrombin-like enzymes is weak or nil. Also it is presently admitted that most of the envenomations involve an afibrinogenaemia, which sometimes is combined with fibrinolysis, but not by a DIC syndrome. Consequently the heparin therapy has lost many of its partisans and its use in the treatment of bites by exotic snakes has diminished considerably.

The inhibiting activity of the antifibrinolytics on the enzymes in viperid venoms has not been explored experimentally. These drugs (aprotinine, ε-amino-caproic acid and tranexamic acid) are rarely used in the treatment of snakebites. It would be of interest to have them evaluated with more precision.

2.5 Treatment of the viper syndrome

The treatment must be considered *in toto*. While the survival of the patient occupies the foreground, the grave risks of severe sequelae linked to the syndrome, must not be neglected. The objective must therefore be to limit the risks of complications, notably necrosis and gangrene, in order to preserve future functionality.

In first aid all aggressive or dangerous gestures must be strictly forbidden, like placing a tight tourniquet, local incisions, the use of physical or chemical agents meant to destroy the venom or delay its diffusion. These heroic treatments are still seen too often and they are probably responsible for most of the sequelae.

In contrast it is essential to disinfect the wound carefully, as soon as possible after the bite, and after that regularly two times a day in an antiseptic solution (quaternary ammonium, Dakin's solution) in a prolonged bath. The preventive use of broad-spectrum antibiotics has its partisans, but its usefulness has never been proven. The therapy with hyperbaric oxygen has given disappointing results, but where it is available, it can prevent or contain gas gangrene.

The removal of tissue debris is often recommended by surgeons, but it does not prevent the necrosis. The value of incisions in the prevention of a Volkmann type complication is limited to cases with a high intra-compartmental pressure. However, the latter is well controlled by an early immunotherapy at elevated doses. Consequently the indication for such a surgical intervention is the abnormal pressure increase within a muscle compartment. This remains a rare eventuality in comparison to the lesions caused by the venom itself and for which the only therapy is the immunotherapy. Without even mentioning the danger of a surgical intervention in the case of a viperid bite, the prevention of necrosis is now practically assured just by an early immunotherapy. Also, for limiting the necrotizing action of the venom, it has to be neutralized well by the antivenin immunoglobulin fragments, and this often needs much higher does than those recommended to prevent the death of the patient. However, the therapeutic protocol has not yet been laid down precisely and is still largely empiric. One must simply think that an excess of antibody in the circulation will take care of the enzymes present in the body.

The symptomatic therapy retains its importance, if only to relieve the patient.

The pain often resists against all classical medication. It may be necessary to use narcotic pain killers or a local/regional anaesthesia, at least in the first few days. The salicylates are contra-indicated because of the risk of hemorrhages. The corticoids, which present similar risks, can still be used if the blood coagulation is controlled, because of their inhibiting effects on many of the enzymes contained in the venoms. The edema can also be combated with the help of diuretics, which in addition will enhance the elimination of the venom. It will be necessary to watch the hydro-electrolyte equilibrium in view of the additions (immunotherapy by perfusion) and the eventual losses. Evidently this will be very difficult to monitor in peripheral health care centers in tropical countries.

The use and local application of enzyme inhibitors, notably EDTA, can be proposed. Their general administration is still in the experimental stage.

The etiological complexity of the viper syndrome, the absence of excessive intra-compartmental pressure, the risks of hemorrhage, and the danger of secondary infection, particularly in the tropics, all argue in favor of a prepared wait and see approach. Drainage incisions or a decompression fasciotomy have never proved themselves and often lead to complications. The excision of necrotic areas is balanced by the need of repeated interventions that are suffered poorly and are demoralizing because of their evident inefficacy. Surgery comes into its own rights after all venom has been eliminated from the body and the lesions have become stabilized, which may take several weeks. Some cases which allow one to suspect a deep muscular lysis (notably bites by *B. brazilii*) can justify an exploration and an eventual drainage of the deep tissue lesions.

2.6 Treatment of the complications

In addition to necrosis, the treatment of which has been discussed above, the two other main complications happening in the course of a viperine envenomation are cerebro-meningial hemorrhage, probably the cause of large part of the deaths, and renal insufficiency.

The first one is difficult to avoid if the hemorrhagic syndrome is not treated properly from the start. It could benefit from a treatment with corticosteroids together with mannitol.

The second one can be prevented by an early stimulation of diuresis and its maintenance at the level of about 50 ml per hour during the whole duration of the envenomation. It is absolutely necessary to carry out regular checks for proteinuria and microscopic haematuria. The treatment of renal insufficiency includes peritoneal dialysis which is the more effective the earlier it is started.

2.7 Monitoring

The patient has to be monitored until his complete recovery. This will comprise clinical examination and biochemical tests, and their thoroughness and precision will depend on the infrastructure and means of the health care center (cf. Figs. 52 and 54).

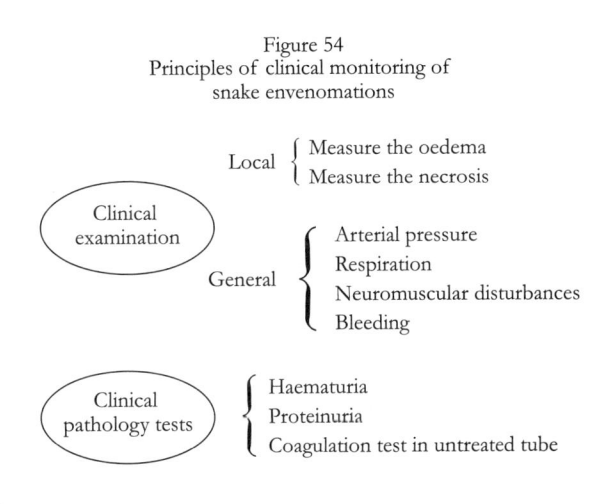

Figure 54
Principles of clinical monitoring of snake envenomations

Clinical examination
- Local: Measure the oedema / Measure the necrosis
- General: Arterial pressure / Respiration / Neuromuscular disturbances / Bleeding

Clinical pathology tests
- Haematuria
- Proteinuria
- Coagulation test in untreated tube

The local developments (edema and necrosis) are monitored twice daily, as well as the clinical and biochemical examination of the nervous (reflexes), the respiratory (rhythm), cardiovascular (arterial pressure) and renal systems (quantity and quality of the urine).

3. Evaluation of the Availability of Antivenins and the Improvement of Their Distribution

Although antivenin is the only specific treatment of envenomations and is very efficient, it is found paradoxically that it is under-used, particularly in those countries which are most in need of it: the developing countries. Recent studies carried out in these countries including in Asia, where the situation is better, though, show that the use of antivenins presently is below 10% of the actual needs, and even 1% in some countries in sub-Saharan Africa. This important difference can have four causes, which are not exclusive, though.

- **An ignorance of the needs**, due to the lack of pertinent epidemiological information and of precise determination of the medical problem;

- **An insufficient supply** because of both too low a quantity produced and a deficient distribution;

- **A usage rate affected by poorly analyzed particularities** like the inability of the health structures to treat snake envenomations, the refusal of certain users to use Western medicine or an opposition to the use of the product by the victim or his family or by the health care personnel;

- **A high price of the product**, the price of which in some cases equals several months' income of a peasant family.

3.1 Analysis of needs, supplies and usage

One has to distinguish between demand and needs, which is defined by the gap between the actual state of health and the desired state of health. The latter consists of a compromise between the aspiration of a population trying to exclude the risk and a reasonable objective in view of the available means, particularly in a developing country.

With regard to antivenin the problem of supply is situated at two levels: the producer and the prescribing physician. The first, contrary to what is normally observed in pharmaceutical distribution, shows an excessive prudence in the distribution of his product. The producer considers primarily the high cost of production of the antivenin, particularly of the highly purified antivenin immunoglobulin fractions, in relation to a restricted and/or insolvent market. The difficulties to capture the needs of the market have led most of the producers to reduce considerably their volume of production in an effort to adapt to a clearly

Clinic and Treatment of Envenomations

diminishing demand. Consequently of the 64 producers, who produced 158 snake antivenins in the 1970s, there were only 22 left in 2000 who sold 73 antivenins.

The prescribing physician on the other hand is subjected to real or supposed constraints, which considerably limit his orders. Amongst the reasons given by the health personnel for limiting their stocks, one can cite the high price of the product, the difficulties of storing it under the climatic conditions in tropical countries and the short expiry time.

The use is influenced by a complex approach linked to the way the public sees envenomations, the therapeutic choices and the appropriateness of the response by the health system. The envenomation is always feared as a natural or accidental disease event. In many communities it is seen as a punishment or vengeance that can be deflected by traditional methods rather than by modern medicines. In many developing countries the initial and sometimes the only help is sought from traditional medicine. This choice often is accentuated by the difficulties to reach a health care center or by the dilapidated state of the health care structures. Finally, the lack of education and the poor availability of health care personnel are the cause of a chronic under-use of the antivenins.

This situation feeds a vicious circle (Fig. 55), which illustrates the complexity of the problem.

Figure 55
The vicious circle feeding the poor availability of antivenoms

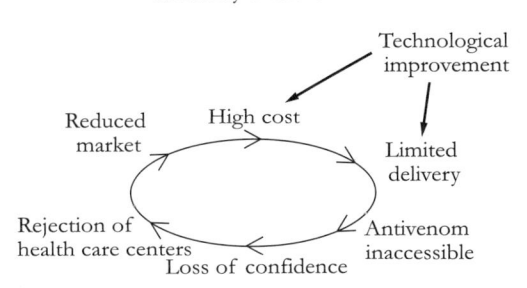

3.2 Improvement of availability and use of antivenins

Epidemiological studies have to be conducted to determine the precise needs for antivenins, and these studies have to be carried out thoroughly and following appropriate methods.

The response to the lack of supply appears to be more complex. Contrary to what is generally proposed it is not the improvement of the product itself that is the cause. The technological level reached today can be considered as sufficient. An effort still has to be made to reduce the price of the antivenin, e.g. by simplifying the presentation and the packaging, or by de-localizing and rationalizing the production. In addition one can propose new ways of financing: subsidies by the state or by communities, participation by agricultural enterprises, private contributions, cost recovery, adherence to the protocol of orphan medicines which consists of supporting research and production, to develop investments and the production of the product and contribute to the training of the health care personnel.

The essential point seems to be the improvement of distribution, particularly in rural areas. The placing of stocks in necessary quantities should correspond to the needs when they have been worked out precisely. It can also be necessary to employ new commercial practices (gratis exchange of expired lots, monitoring the distributors, price fixing).

The training of the health care personnel with a view to improving the admission and treatment of the patient remains essential.

4. Prevention of Envenomations

The prevention aims on one hand to reduce the number of snakebites and on the other hand the severity of the envenomations and the mortality. It can be organized at three levels. One can control the snakes, prevent the bites or at least reduce their frequency, and organize the admission of envenomation cases, which should be an objective of priority in developing countries.

4.1 Snake control

For the control of snake populations several strategies have been proposed with varying success. In any case it should be remembered that the snakes contribute to the balance of the environment and that their control can cause a multiplication of their prey animals, notably rodents, which are equally pests.

4.1.1 Territorial protection and control by physical means

A snake capture policy encouraged by a premium was practiced in Brazil at the beginning of the last century with a moderate success; in 1926 it allowed the capture of 120,000 snakes, which is little for the large country, and even more so as there is no record of the identification of the collected specimens. In Martinique this strategy was employed against *Bothrops lanceolatus*, the only venomous snake on the island. It was soon abandoned again, more likely because of its inefficacy than because it reportedly led to fraudulent practices consisting in rearing the females with the aim to produce the many young ones born each year: a female can give birth to up to fifty young ones. The remarkable fecundity of the *Bothrops* spp. and their ecological tolerance alone can explain the failure.

In certain North American metropoles the firemen are required to collect and dispose of snakes reported by the inhabitants. This method was supposed to have the advantage of protecting both people and animals… Not only is such a strategy expensive, its efficacy has not been confirmed either.

Keeping dogs trained to detect snakes has been proposed in Japan, and similarly one has used poultry (chickens, turkeys, ducks) to announce the presence of reptiles.

The installation of protective devices (ditches with smooth sides around a plantation, nylon netting) has been used experimentally in Japan. The use of electrified barriers has allowed one to reduce the snake density by 50% in the fenced-in area. Trapping requires a large number of traps and their methodical distribution: on the Anami islands in the South of Japan it was calculated that that 200 traps per hectare were needed to eradicate the snakes within 6 months. However, these techniques are costly to install and to run and need particular topographical conditions to be really effective.

4.1.2 Ecological control

The modification of the environment has an effect on the snake population, either by favoring the development of certain populations or on the contrary by limiting their conservation.

The creation of large fields has enabled an extensive and heavily mechanized agriculture and has caused a decrease in snakebite morbidity in many industrialized countries.

The destruction of certain habitats (pack walls, elimination of scrub growth, wind slash) allows one to reduce the density of snake populations in certain areas. This method appears to have given excellent results on the tropical islands of Japan.

4.1.3 Biological control

Mongooses, notably the Indian mongoose (*Herpetes edwardsi*), a famous predator of snakes, have been introduced in several tropical regions to fight snakes. Their broad carnivorous habits have been confirmed on the Anami islands, in Trinidad and Puerto Rico, where even a relative competition could be observed between certain venomous snakes and the mongooses. However, snakes constitute less than 10% of the prey captured by the mongooses. In addition the mongooses hunt mainly by day and their encounters with venomous snakes, which often are nocturnal, are largely governed by chance. Finally it can be argued that the mongoose affects equally the whole of the vertebrate fauna and therefore can adversely affect the environment.

Several researchers have experimented with the impact of parasitic, bacterial or viral infections on snake populations. The hemogregarines (*Haemogregarina* spp. or *Hepatozoon* spp.) are blood parasites of reptiles and infect more than half of the snakes captured in the primary equtorial forest without having any visible effect on their health or abundance. The amoeba (*Entamoeba invadens*) is a well-known pathogen, but it is difficult to infect the snakes without endangering the non-target fauna.

4.1.4 Chemical control

This strategy has been proposed for a long time already but with only modest success.

The use of poisoned water or prey animals, notably with nicotine or strychnine, has given interesting results in North America and in Japan before the 1960s. Undesirable effects on the environment and the dangers of such practices have led to them being abandoned.

The use of repellents or of pesticides - organochlorines, combinations of naphthaline and sulfur (Snake-A-Way®, traded in the USA), pyrethroids (deltamethrine, an insecticide toxic to cold blooded animals: K-Othrine®, Décis®) – necessitate high doses which is costly and dangerous for the environment. The fumigation with methyl bromide is effective but brings with it the considerable danger of destroying the ozone layer. Certain plants appear to be able to keep snakes away without danger to other vertebrates. *Securidaca longepedunculata*, a famous African anti-venomous plant, and *Ipomoea carnea*, an American shrub occurring from Texas to Argentina, are said to be strong natural repellents. Recently a product of unspecified composition is being traded in Europe as repellent specifically against snakes (Vivert®).

4.2 Protection against bites

One of the simplest and least onerous means is protective clothing. Wearing heavy boots and thick gloves reduces significantly the danger of snakebite. Wide-brimmed hats protect against arboreal snakes that can fall onto the victim and bite on the head. It is necessary to avoid certain dangerous actions or to take precautions beforehand: sounding cavities with a stick, using a tool or protecting the hand. Finally the mechanization of agriculture is also an important factor in risk reduction by limiting the direct man/snake contact.

The individual prevention rests largely on wide-ranging information of the public on the circumstances of accidents, the ecology of the dangerous species and the first aid to be given in the case of a bite. This advice is given out in schools, via the press or as posters. Although it has not truly shown its efficacy this method has the merit of being inexpensive.

4.3 Organization of the admission to medical care

The speed of intervention and of starting the correct treatment is the essential factor.

It is necessary to avoid panic, which often leads to detrimental actions. The first aid has to be simple and this requires adequate information of the public who should be induced to consult a health care center as soon as possible. Naturally, the organization of the health system should be such as to meet the expectations of the public and its needs.

The response given to snake envenomations should not only be scientific but also pragmatic and operational. The fact that in spite of their efficacy the antivenins are under-utilized has as consequence that morbidity and lethality remain at an unacceptable level and this leads to focusing again the analysis that has been carried out on this problem.

Two major problems become apparent. The unavailability of the antivenin for which the producer, the prescribing physician and the beneficiary share the responsibility. The main reason appears to be the price of the product, which is higher than the average buying power of the patients. The other one is that the use of the product remains deficient without a precise protocol and adequate training.

Finally, a daring marketing, financial and health education policy could improve the situation and lower mortality significantly.

APPENDIX 1

Practical Measuring of the LD$_{50}$ by the Spearman-Kärber Method

Conditions:

- Constant value of d (= logarithm of the realation between 2 successive doses)
- covering all the response intervals between 0% and 100%
- symmetrical distribution of the responses
- at least 5 animals for each dilution
- observation during 48 hours (some stipulate 24 hours)
- 5 control animals (injection of diluent without venom)

Technique:

- prepare the venom dilutions starting from the lowest to the highest concentration
- inject all the mice with the doses by the same route, in the same volume and during the same manipulation
- count the dead mice after the required delay
- apply the following formula:

$$\log LD_{50} = x_{100} \pm \frac{d}{n} (\Sigma \, r - n/2)$$

- the variance equals:

$$V(m) = \frac{d^2}{n^2(n-1)} \Sigma \, [r(n-r)]$$

where x_{100} is the logarithm of the lowest dose causing 100% mortality
n is the number of animals used for each dilution
r is the number of dead animals for each dilution
t = 2.20 for p = 0.05 with d.d.l. = Σ (n-1) including all the groups of animals falling between x0 et x100, these two values being excluded

Example:

Dose of venom (mg)	0.28	0.35	0.44	0.55	0.69	0.86	1.07
Number of deaths	0/5	1/5	1/5	3/5	4/5	5/5	5/5

$d = \log 1.25 = 0.223$
$x_{100} = \log 0.86 = -0.151$
$x_0 = \log 0.28 = -1.273$
$n = 5$
d.d.l. = 6

$$LD_{50} = -0.151 \pm \frac{0.223}{5} [0+1+1+3+4+5-(5/2)] = -0.151 \pm \frac{0.223}{5} (11.5) = -0.151 \pm 0.513$$

the two possible values for the LD_{50} are antilog (-0.664) and antilog (0.362), the latter value being by definition inappropriate.

$$V(m) = \frac{0.223^2}{5^2(5-1)} [(0\times5)+(1\times4)+(1\times4)+(3\times2)+(4\times0)] = \frac{0.0497}{100} \cdot 14 = 0.00696$$

Consequently the LD_{50} is 0.51 mg (IC = 0.43-0.62)

APPENDIX 2

Symptomatic Action of Certain Plants

(C) = cataplasma; (D) = decoction; (I) = infusion; (M) = maceration; (?) = not specified

Plant	Analgesic	Anti-inflammatory	Antiseptic	Cicatrizing	Haemostatic	Diuretic	Stimulant	Sedative	Specific
Acacia albida (ana tree)			bark (C)	root (C)			bark		
Acacia dudgeoni			root	leaf					root
Acacia erythrocalyx									?
Acacia macrostacya		bark							leaf
Acacia nilotica				fruit (C)	fruit (C)	root (I)			
Acacia polyacantha		root[1], bark[1],							root
Acacia sieberiana		root[1]			bark				bark
Achyranthes aspera						plant (D)			
Adansonia digitata (baobab tree)	leaf, seed	bark, leaf	gum	fruit		leaf	seed[4]		
Agave sisalana (agava)						leaf			
Ageratum conyzoides			leaf (C)						
Aloe vera (aloe)		gel		gel	gel[2]				
Anacardium occidentale		bark	plant	stem, fruit[5]					
Ananas comosus (pineapple)		stem, fruit[3]				fruit	fruit	leaf	root, bark
Annona senegalensis				gum					
Aphania senegalensis				leaf					bark
Azadirachta indica (neem tree)		bark, leaf (D)	young leaf, seed (?)			bark (D)	root, flower		
Balanites aegyptica		fruit (C)[1]					leaf		root
Boerhavia diffusa					root[6]	root			

[1]Anti-oedematous; [2]Anti-platelet-aggregation; [3]Plasmine-like = thrombolytic; [4]Cardiotonic; [5]Anti-nécrotic (proteolysis); [6]Antifibrinolytic

(C) = cataplasma; (D) = decoction; (I) = infusion; (M) = maceration; (?) = not specified

Plant	Analgesic	Anti-inflammatory	Antiseptic	Cicatrizing	Haemostatic	Diuretic	Stimulant	Sedative	Specific
Borassus aethiopum				flower					
Borreria verticillata			leaf (C)	leaf (C)		root (M)			
Boswellia dalzielli									root + bark
Bridelia micrantha		bark + leaf	root, fruit branch						
Butyrospermum parkii (shea)		butter (C)		butter (C)	butter (C)				seed
Calotropis procera						root	bark	?	
Capparis tomentosa	root					?			root, leaf
Capsicum frutescens (pili pili)		fruit[2]							
Carica papaya (pawpaw tree)			seed (C)	latex (C)	fruit (C),	root, leaf latex (C)			
Cassia alata		branch[1]	leaf (C)						
Cassia occidentalis			seed (C)			root, leaf (M, D)	root, seed		
Cassia sieberiana			seed (C)			root, leaf (M, D)		root, seed	
Casuarina equisetifolia	?								
Celtis integrifolia		leaf	leaf	leaf					
Centella asiatica				leaf (C)		?			
Chrysanthelum indicum					plant (D)				
Cissus populnea		root[1], stem							leaf (ash)

[1]Anti-oedematous; [2]Anti-platelet-aggregation; [3]Plasmine-like = thrombolytic; [4]Cardiotonic; [5]Anti-nécrotic (proteolysis); [6]Antifibrinolytic

(C) = cataplasma; (D) = decoction; (I) = infusion; (M) = maceration; (?) = not specified

Plant	Analgesic	Anti-inflammatory	Antiseptic	Cicatrizing	Haemostatic	Diuretic	Stimulant	Sedative	Specific
Cochlospermum tinctorium		root (D)			?	?			?
Cola nitida (cola tree)							nut		
Combretum sp.		leaf² (D)	leaf (D)	seed (C)	leaf (D)	leaf (D)			
Commiphora africana (hairy corkwood)	leaf, bark		resin	leaf, bark					
Costus afer	root (D)	leaf							
Crateva adansonii		root¹ (C), bark							
Crescentia cujete				leaf, fruit	leaf			fruit	nut (ash)
Crossopteryx febrifuga	root	root						root	
Daniellia oliveri				bark, stipulate	bark	root			bark
Datura metel		leaf (C)						leaf	
Dichrostachys cinerea (sickle bush)		bark (I)				leaf, root			leaf, fruit, bark
Dombeya quinqueseta									bark
Entada africana							root		
Erythrina senegalensis (coral tree)	bark, leaf, nut			leaf, root					
Erythrophleum suaveolens	bark								bark
Euphorbia balsamifera	latex (C)		latex (C), branch	latex					latex, branch
Euphorbia tirucalli			latex						root

¹Anti-oedematous; ²Anti-platelet-aggregation; ³Plasmine-like = thrombolytic; ⁴Cardiotonic; ⁵Anti-nécrotic (proteolysis); ⁶Antifibrinolytic

(C) = cataplasma; (D) = decoction; (I) = infusion; (M) = maceration; (?) = not specified

Plant	Analgesic	Anti-inflammatory	Antiseptic	Cicatrizing	Haemostatic	Diuretic	Stimulant	Sedative	Specific
Feretia apodanthera									root, leaf, flower
Ficus capensis (Cape fig)		root[1]				?			
Ficus natalensis								root	root
Ficus sycomorus	latex								leaf
Ficus thonningii				bark (C)					
Gardenia ternifolia		bark[2]		bark (C)		root	bark		?
Grewia bicolor						bark		bark	root
Guiera senegalensis	leaf, root (D)	leaf[3], root (D)	leaf (C)	leaf (C)	leaf (C)	leaf (I), gall			
Hibiscus esculentus			flower (C)			flower, leaf (D, I)	flower (D, I)		
Hibiscus sabdariffa						bark, root, leaf (D,M)			
Holarrhena floribunda					leaf (C)				
Hoslundia opposite		leaf	leaf	root		leaf			leaf (prevention & treatment)
Jatropha curcas (physic nut)	leaf	leaf[4]		latex	latex				latex
Kalanchoe pinnata		leaf (C)	leaf (C)						
Khaya senegalensis	root, bark	root, bark	root, stalk		root, bark				
Kigelia africa				bark					root

[1]Anti-oedematous; [2]Anti-platelet-aggregation; [3]Plasmine-like = thrombolytic; [4]Cardiotonic; [5]Anti-nécrotic (proteolysis); [6]Antifibrinolytic

(C) = cataplasma; (D) = decoction; (I) = infusion; (M) = maceration; (?) = not specified

Plant	Analgesic	Anti-inflammatory	Antiseptic	Cicatrizing	Haemostatic	Diuretic	Stimulant	Sedative	Specific
Lannea schimperi									bark
Lausonia inermis (henna)			leaf (C)	leaf (C)	leaf (C)	root			
Leptadenia hastata				latex					flower
Macrosphyra longistyla					bark	root		flower	
Mangifera indica (mango tree)					leaf (D, I), core	leaf (D, I)			
Maytenus senegalensis		leaf, bark		root, bark			bark		root
Mimosa pigra				bark			root (I)		root
Mitragyna inermis (ben gum)						leaf, bark			
Moringa oleifera	root (C)		plant	root (C)		gum, leaf	gum⁴		leaf
Olax subcorpioides									leaf, bark
Parinari curatellifolia			bark	bark					branch
Parkia biglobosa			bark, leaf						bark
Pavetta crassipes									bark, leaf
Pericopsis laxiflora	bark, leaf								
Piliostigma reticulatum		bark, leaf (C)		bark, leaf (C)	bark, leaf (C)			leaf	
Piliostigma thonningii				fruit					leaf
Psidium guajava (guava tree)			leaf (C)	bark = + gum		branch + leaf			
Psorospermum senegalense								leaf	root

[1]Anti-œdematous; [2]Anti-platelet-aggregation; [3]Plasmine-like = thrombolytic; [4]Cardiotonic; [5]Anti-nécrotic (proteolysis); [6]Antifibrinolytic

(C) = cataplasma; (D) = decoction; (I) = infusion; (M) = maceration; (?) = not specified

Plant	Analgesic	Anti-inflammatory	Antiseptic	Cicatrizing	Haemostatic	Diuretic	Stimulant	Sedative	Specific
Ricinus communis (ricinus)	leaf (C)	leaf (C)	nut	bark	nut				
Saba comorensis					latex				bark
Sarcocephalus latifolius		root, leaf			root, leaf	leaf			bark
Sclerocarya birrea	branch	bark					nut		bark
Securidaca longepedunculata	root (C)	root (C)						root (D)	root, branch, leaf
Smilax anceps						root			leaf
Sterculia setigera						root			veterinarian use
Stereospermum kunthianum				bark, leaf	bark	root	leaf		bark
Strychnos innocua	root, leaf	root							root
Strychnos spinosa (monkey orange)	leaf		root	root		root		root	leaf
Suartzia madagascarensis									root
Tabernanthe iboga							root		
Tamarindus indica (tamarind)		fruit, leaf (C)	leaf (C)	bark (C)		bark	bark, branch		
Terminalia avicennioides		root, bark	root, bark	root, bark	root, bark				
Terminalia macroptera				root, bark	root (C), bark	leaf, root	root		bark
Tetracera alnifolia								branch	root, leaf

[1] Anti-œdematous; [2] Anti-platelet-aggregation; [3] Plasmine-like = thrombolytic; [4] Cardiotonic; [5] Anti-nécrotic (proteolysis); [6] Antifibrinolytic

(C) = cataplasma; (D) = decoction; (I) = infusion; (M) = maceration; (?) = not specified

Plant	Analgesic	Anti-inflammatory	Antiseptic	Cicatrizing	Haemostatic	Diuretic	Stimulant	Sedative	Specific
Thevetia neriifolia							nut, fruit		root
Tinospora bakis		root[1]				root			root
Trema guineensis	leaf	leaf (D)				leaf (D)			branch, leaf
Uvaria chamae				leaf					
Vetiveria nigritana (vetiver)			root (C)	root (C)	root, bark, leaf				
Vitex doniata			bark, leaf, fruit	leaf (C)			leaf (D), bark		veteranerian use
Vitex simpecifolia			bark						root (ash)
Ximenia americana (sour plum)		root[1]	root (C)	leaf (C)	leaf, root (C)				
Xylopia aethiopica (bitterwood)			bark, fruit (C)						
Zanthoxylum zanthoxyloides	leaf	root, bark[1]		root, leaf					root
Ziziphus mauritania (jujube)		fruit, leaf[6]			bark (C)	fruit			
Ziziphus sp-na-christi				spine (C), leaf				leaf	spine

[1]Anti-oedematous; [2]Anti-platelet-aggregation; [3]Plasmine-like = thrombolytic; [4]Cardiotonic; [5]Anti-nécrotic (proteolysis); [6]Antifibrinolytic

APPENDIX 3

Antivenom Producers and Their Products

Europe

Producer	Name of product	Neutralised venom	Animal used	Method of purification	Description and presentation of the product
Aventis Pasteur 2 Avenue Pont Pasteur, 69367 Lyon cedex 07, FRANCE	Viperfav	*Vipera aspis*, *V. berus*, *V. ammodytes*	Horse	Precipitation Peptic digestion Adsorption on gel Chromatography	In liquid form 2 ml syringe 1 ml neutralises 250 LD_{50} de *V. aspis* and *V. ammodytes* and 125 LD_{50} of *V. berus*
	Favirept	*Bitis arietans* *Echis leucogaster* *Naja haje*, *N. nigricollis* *Cerastes cerastes* *Macrovipera deserti*	Horse	Precipitation Peptic digestion Adsorption on gel Chromatography	In liquid form 10 ml ampule 1 ml neutralises 25 LD_{50} of *B. arietans*, *E. leucogaster* and *N. haje* and 20 LD_{50} of the other venoms
	Favafrique	*Bitis arietans*, *B. gabonica*, *B. nasicornis* *Echis leucogaster*, *E. ocellatus* *Naja haje*, *N. melanoleuca*, *N. nigricollis* *Dendroaspis jamesoni*, *D. polylepis*, *D. viridis*	Horse	Précipitation Digestion pepsique Adsorption on gel Chromatography	In liquid form 10 ml ampule 1 ml neutralises 25 LD_{50} of *Bitis*, *Echis*, *Dendroaspis* and *N. haje* and 20 LD_{50} of *N. melanoleuca* and *N. nigricollis*
	Bothrofav	*Bothrops lanceolatus*	Horse	Precipitation Peptic digestion Adsorption on gel Chromatography	In liquid form 10 ml ampule 1 ml neutralise 25 LD_{50}

Europe (continued)

Producer	Name of product	Neutralised venom	Animal used	Method of purification	Description and presentation of the product
Institute of Immunology, Inc, Rockefeller str. 2, Zagreb, CROATIA	European Viper Venin Antiserum	*Vipera ammodytes, V. aspis, V. berus, V. lebetina, V. xanthina, V. ursinii*	Horse	Precipitation Peptic digestion	In liquid form 10 ml ampule 1 ml neutralises 100 LD_{50} of *V. ammodytes* and *V. aspis*, 50 LD_{50} of the venoms of the other vipers.
Knoll AG, Postfach 21 08 05, 67008 Ludwigshafen, GERMANY	Snake Antivenom	*Naja naja sputatrix*	Goat	Precipitation Peptic digestion Purification	In liquid form 1 ml contains 100 mg of proteins max. 10 ml ampule

Asia

Producer	Name of product	Neutralised venom	Animal used	Method of purification	Description and presentation of the product
National Antivenom and Vaccine Production Centre, National Guard Health Affairs, Riyadh, Kingdom of SAUDI ARABIA	Polyvalent Snake Antivenom	*Bitis arietans Cerastes cerastes Echis carinatus, E. coloratus Naja haje Walterinnesia aegyptia*	Horse or goat	Precipitation Peptic digestion Adsorption on gel	In liquid form; protein concentration between 1 and 1.3 g 1 ml neutralises at least 25 LD_{50} 10 ml ampule
	Bivalent Snake Antivenom	*Naja haje Walterinnesia aegyptia*	Horse	Precipitation Peptic digestion Adsorption on gel	In liquid form; protein concentration 1 g 1 ml neutralises at least 100 LD_{50} 10 ml ampule
	Bivalent Atractaspis/ Walterinnesia	*Walterinnesia aegyptia truataspis microlepidota, A. engaddensis*	Horse	Precipitation Peptic digestion Adsorption on gel	In liquid form; protein concentration 1g% 1 ml neutralises at least 25 LD_{50} *Walterinnesia Atractaspis* and 200 LD_{50} *Walterinnesia* 10 ml ampule

Asia (continued)

Producer	Name of product	Neutralised venom	Animal used	Method of purification	Description and presentation of the product
Razi Vaccine & Serum Research Institute, P.O. Box 1365, 1558 Tehran, IRAN	Antisnake Bite Serum	Naja naja oxiana, Macrovipera lebetina, V. albicornuta, Echis carinatus, Pseudocerastes persicus, Agkistrodon halys	Horse	Precipitation Peptic digestion	In liquid form 10 ml ampule
Razi Vaccine & Serum Research Institute, P.O. Box 11365, 1558 Tehran, IRAN	Antisnake Bite Serum (monovalent)	Naja naja oxiana	Horse	Precipitation Peptic digestion	In liquid form 10 ml ampule
	Antisnake Bite Serum (monovalent)	Macrovipera lebetina	Horse	Precipitation Peptic digestion	In liquid form 10 ml ampule
	Antisnake Bite Serum (monovalent)	Echis carinatus	Horse	Precipitation Peptic digestion	In liquid form 10 ml ampule
PT Bio Farma, Jl. Pasteur 28, P.O. Box 47, Bandung 40161, INDONESIA	Polyvalent anti snake Venom	Calloselasma rhodostoma, Bungarus fasciatus, Naja sputatrix	Horse	Precipitation	In liquid form 5 ml ampule

Asia (continued)

Producer	Name of product	Neutralised venom	Animal used	Method of purification	Description and presentation of the product
Thai Government Pharmaceutical Organization, 75/1 Rama VI Rd, Bangkok 10400, THAILAND	Cobra Antivenom	*Naja naja kaouthia*	Horse	Precipitation Enzyme digestion	In liquid form 1 ml neutralises 0.6 mg of venom
	Malayan Pit Viper Antivenom	*Calloselasma rhodostoma*	Horse	Precipitation Enzyme digestion	In liquid form 1 ml neutralises 1.6 mg of venom
	Russell's Viper Antivenom	*Daboia russelli*	Horse	Precipitation Enzyme digestion	In liquid form 1 ml neutralises 0.6 mg of venom
Central Research Institute, Simla Hills, H. P. Kasauli, INDIA	Polyvalent Antisnake Venom Serum, IP	*Naja naja* *Bungarus caeruleus* *Daboia russelli* *Echis carinatus*	Mule/Horse	Precipitation Peptic digestion Filtration	In liquid and in lyophilized form 1 ml neutralises 0.6 mg of *Naja* & *Vipera*, 0.45 mg of *Bungarus* & *Echis*
Haffkine Bio-Pharmaceutical Corpn. Ltd, Acharya Donde Marg, Parel, Mumbai 400012 INDIA	Snake Antivenom IP	*Naja naja* *Daboia russellii* *Bungarus caeruleus* *Echis carinatus*	Horse	Enzyme digestion Precipitation Dialysis	In liquid and in lyophilized form (10 ml)
Serum Institute of India Ltd, 212/2, Hadapsar, Pune – 411 028 INDIA	Snake Antivenin Serum IP SII Polyvalent Anti-Snake Venom Serum	*Naja naja* *Bungarus caeruleus* *Daboia russelli* *Echis carinatus*	Horse	Precipitation Diafiltration	In lyophilized form (equivalent of 10 ml) 1 ml neutralises 0,6 mg of *Naja* & *Vipera*, 0,45 mg of *Bungarus* & *Echis*
	Snake Venom Anti-sera	*Dendroaspis polylepis* *Bitis gabonica* *Daboia russelli* *Echis carinatus*	Horse	Precipitation Diafiltration	In lyophilized (equivalent of 10 ml) 1 ml neutralises 0.6 mg of *Bitis* & *Vipera*, 0.45 mg of *Dendroaspis* & *Echis*

Asia (continued)

Producer	Name of product	Neutralised venom	Animal used	Method of purification	Description and presentation of the product
Bharat Serums & vaccines Ltd 16th floor, Hoechst House, Nariman Point, Mumbai – 400 021 INDIA	Anti snake venom serum Asia	*Daboia russelii Echis carinatus Naja naja Bungarus caeruleus*	Horse	Precipitation Enzyme digestion Diafiltration	In liquid form; protein concentration between 1 and 2 g 1 ml neutralises 0.6 mg of *Naja* & *Daboia*, 0.45 mg of *Bungarus* & *Echis* 10 ml ampule
	Anti snake venom serum Africa	*Bitis gabonica, B. arietans, B. nasicornis Dendroaspis jamesonii, D. polylepis, D. angusticeps Echis carinatus, E. ocellatus Naja nivea, N. nigricollis, N. annulifera, N. melanoleuca*	Horse	Precipitation Enzyme digestion Diafiltration	In liquid form; protein concentration between 1 and 2 g 1 ml neutralises at least 25 LD$_{50}$ 10 ml ampule
Biologicals Production Service, Filinvest Corporate City, Alabang, Muntinlupa, PHILIPPINES	Cobra Antivenom	*Naja naja philippinensis*	Horse	Precipitation Peptic digestion Filtration on membrane	In liquid form

Asia (continued)

Producer	Name of product	Neutralised venom	Animal used	Method of purification	Description and presentation of the product
Shanghai Institute of Biological Products, 1262 Yang An Road (W., Shanghai 200052, CHINA	Purified Agkistrodon Acutus Antivenom	*Deinagkistrodon acutus*	Horse	Precipitation Peptic digestion Filtration on gel	In liquid form
	Purified Agkistrodon Halys Antivenom	*Gloydius halys Protobothrops mucrosquamatus, Trimeresurus stejnegeri*	Horse	Precipitation Peptic digestion Filtration on gel	In liquid form
	Purified Bungarus Multicinctus Antiv.	*Bungarus multicinctus*	Horse	Precipitation Peptic digestion Filtration on gel	In liquid form
	Purified Naja Antivenom	*Ophiophagus hannah*	Horse	Precipitation Peptic digestion Filtration on gel	In liquid form
	Purified Vipera Russelli Antivenom	*Daboia russelli*	Horse	Precipitation Peptic digestion Filtration on gel	In liquid form
Takeda Chemical Industries Ltd, Osaka, JAPAN	Freeze-Dried Mamushi Antivenom	*Gloydius halys*	Horse	Not available	In lyophilized form

Asia (continued)

Producer	Name of product	Neutralised venom	Animal used	Method of purification	Description and presentation of the product
The Thai Red Cross Society, Queen Saovabha Memorial Institute, Bangkok, THAILAND	Cobra Antivenom	*Naja naja kaouthia*	Horse	Precipitation Enzymatic digestion	In lyophilized form (equivalent of 10 ml)
	King Cobra Antivenom	*Ophiophagus hannah*	Horse	Precipitation Enzymatic digestion	In lyophilized form (equivalent of 10 ml)
	Banded Krait Antivenom	*Bungarus fasciatus*	Horse	Precipitation Enzymatic digestion	In lyophilized form (equivalent of 10 ml)
	Russell's Viper Antivenom	*Daboia russelli siamensis*	Horse	Precipitation Enzymatic digestion	In lyophilized form (equivalent of 10 ml)
	Malayan Pit Viper Antivenom	*Calloselasma rhodostoma*	Horse	Precipitation Enzymatic digestion	In lyophilized form (equivalent of 10 ml)
	Green Pit Viper Antivenom	*Trimeresurus albolabris*	Horse	Precipitation	In lyophilized form (equivalent of Enzymatic digestion 10 ml)
Institut Pasteur d'Algérie Rue du Docteur- Laveran, Alger, ALGERIA	Antiviper Serum	*Cerastes cerastes* *Macrovipera lebetina*	Horse	Precipitation Peptic digestion Filtration & dialysis	In liquid form 1 ml neutralises 40 LD_{50} of *Vipera* venom and 20 LD_{50} of *Cerastes* venom 10 ml ampule

Asia (continued)

Producer	Name of product	Neutralised venom	Animal used	Methode of purification	Description and presentation of the product
South African Vaccine Producers (Pty) Ltd, PO Box 28999, Sandringham 2131, SOUTH AFRICA	Polyvalent Snake Antivenom	*Bitis arietans, B. gabonica Hemachatus haemachatus Dendroaspis angusticeps, D. jamesoni, D. polylepis Naja nivea, N. melanoleuca, N. annulifera, N. mossambica*	Horse	Precipitation Heat (56°) Peptic digestion Dialysis Ion exchange filtration/ultration	In liquid form 10 ml ampule
	Boomslang Antivenom	*Dispholidus typus*	Horse	Precipitation Heat (56°) Peptic digestion Dialysis Ion exchange filtration/ultration	In liquid form 10 ml ampule
	Echis Antivenom	*Echis ocellatus*	Horse	Precipitation Heat (56°) Peptic digestion Dialysis Ion exchange filtration/ultration	No stocks
See Aventis Pasteur (Europe)					
See Serum Institute of India (Asia)					
See Bharat lab (Asia					

Americas

Producer	Name of product	Neutralised venom	Animal used	Method of purification	Description and presentation of the product
Wyeth Laboratories Inc. Marletta, PA 17547 USA	North American Coral Snake Antivenom	*Micrurus fulvius*	Horse		In lyophilized form (equivalent of 10 ml) 1 ml neutralises 2 mg or 25 LD$_{50}$ of *M. fulvius* venom
	Polyvalent Antivenom	*Crotalus adamanteus* *C. atrox* *C. durissus terrificus* *Bothrops atrox*	Horse		In lyophilized form (equivalent of 10 ml) High cross-reactivity for the venoms of American and Asian crotalids.
Therapeutic Antibodies Inc. 1207 17th Avenue South, Suite 103, Nashville, TN 37212 USA	Vipera Tab	*Vipera berus* Cross reaction with *V. aspis* & *V. ammodytes*	Sheep	Precipitation Heat (56°) Trypsin digestion Affinity chromatography	In lyophilized form Ampules containing 100 mg of proteins
	EchiTAb	*Echis ocellatus*	Sheep	Idem but Ion exchange chromatography	In lyophilized form
	CroTAb	*Crotalus atrox*, *C. adamanteus*, *C. scutulatus* & other North-American crotalids	Sheep	Idem but Affinity chromatographie	In lyophilized form Ampules containing 500 mg of proteins
	ProlongaTAb	*Daboia russelli*	Sheep	Idem but Ion exchange chromatography	In lyophilized form

Americas (continued)

Producer	Name of product	Neutralised venom	Animal used	Method of purification	Description and presentation of the product
Gerencia Gª de Biologicos y Reactivos, Secretaria de Salud, Amores 1240, Colonia del Valle, Mexico 03100, D.F. MEXICO	Suero Antiviperino polivalente (polyvalent anti-viper serum)	*Crotalus basiliscus* *Bothrops asper*	Horse	Precipitation	In lyophilized form (equivalent 10 ml)
Laboratorios Silanes S.A. de C.V. Amores #1304, Col. Del Valle C.P. 03100 México, DF MEXICO	Antivipmyn	*Crotalus terrificus* *Bothrops atrox*	Horse	Precipitation Peptic digestion	In lyophilized form (equivalent 10 ml) 1 ampule neutralises 780 LD_{50} of *Bothrops* and 200 DL_{50} of *Crotalus* Cross-reactivity with *Agkistrodon* and *Sistrurus*
	Coralmyn	*Micrurus* sp.	Horse	Idem	In lyophilised form (equivalent 10 ml) 1 ampule neutralises 450 LD_{50} of *Micrurus*
Instituto Clodomiro Picado, Univ. De Costa Rica, Ciudad Universitaria "Rodrigo Facio", San José, COSTA RICA	Polyvalent Antivenom	*Bothrops asper* *Crotalus durissus* *Lachesis muta*, *L. stenophys*, *L. melanocephala*	Horse	Fractionation by caprylic acid	In liquid and in lyophilized form (equivalent of 10 ml)
	Anticoral Antivenom	*Micrurus nigrocinctus*, *M. fulvius*, *M. dumerilii*, *M. browni*	Horse	Fractionation by caprylic acid	In liquid and in lyophilized form (equivalent of 10 ml)
	Anti-M. Mipartitus Antivenom	*Micrurus mipartitus*	Horse	Fractionation by caprylic acid	In liquid and in lyophilized form (equivalent of 10 ml)

Americas (continued)

Producer	Name of product	Neutralised venom	Animal used	Methode of purification	Description and presentation of the product
Intituto Butantan, Avenue Vital Brasil, 1500 Sao Paulo, BRAZIL	Soro Antibotropico	Principales espèces de *Bothrops* (*B. atrox, B. cotiara, B. jararaca, B. jararacussu, B. moojeni, B. neuwiedi*)	Horse	Precipitation Peptic digestion	In liquid form 10 ml ampule 1 ml neutralises 5 mg of the reference venom
	Soro Anticrotalico (anti-Crotalus serum)	*Crotalus durissus, C. terrificus, C. collineatus, C. cascavella*	Horse	Precipitation Peptic digestion	In liquid form 10 ml ampule 1 ml neutralises 1.5 mg of the reference venom
	Soro Antielapidico (anti-elapid serum)	*Micrurus corallinus, M. frontalis, M. ibiboboca, M. lemniscatus, M. spixii*	Horse	Precipitation Peptic digestion	In liquid form 10 ml ampule 1 ml neutralises 1.5 mg of the reference venom
	Soro Antibotropico-Crotalico (anti-Bothrops & Crotalus serum)	Cf. Antibotropico and Anticrotalico	Horse	Precipitation Peptic digestion	In liquid form ; 10 ml ampule 1 ml neutralises 5 mg of *Bothrops* and 1,5 mg of *Crotalus*
	Antibotropico-Laquético (anti-Bothrops & Lachesis serum)	Cf. Antibotropico and *Lachesis muta*	Horse	Precipitation Peptic digestion	In liquid form; 10 ml ampule 1 ml neutralises 5 mg of *Bothrops* and 3 mg of *Lachesis*

Australia

Producer	Name of product	Neutralised venom	Animal used	Methode of purification	Description and presentation of the product
Commonwealth Serum Laboratories Ltd. Parkville, Melbourne AUSTRALIA	Brown snake	*Pseudonaja* sp.	Horse	Precipitation, Trypsine digestion	In lyophilized form (equivalent of 10 ml)
	Tiger Snake	*Notechis* sp., *Tropidechis* sp., *Austrelaps* sp. and *Pseudechis* sp.	Horse	Precipitation, Trypsine digestion	In lyophilized form (equivalent of 10 ml)
	Black Snake	*Pseudechis australis*, *P. butleri*	Horse	Precipitation, Trypsine digestion	In lyophilized form (equivalent of 10 ml)
	Taipan	*Oxyuranus* sp.	Horse	Precipitation,	In lyophilized form (equivalent of 10 ml)
	Death Adder	*Acantophis* sp.	Horse	Precipitation,	In lyophilized form (equivalent of 10 ml)

APPENDIX 4

French Legislation Regarding the Keeping of Venomous Animals

General Conditions

The civil responsibility of the owner or guardian of the animal, whichever is the case, implying an obligation of compensation, is engaged as soon as a third party has been prejudiced. The penal responsibility leading to a conviction in the form of a fine or imprisonment can also be invoked if a contravention has been committed. The prejudice may not only result from a damage; however the plaintiff will have to provide proof of prejudice, which can also be moral or psychological. Finally the act may not be intentional: it could be the case of maladroitness, negligence or imprudence.

Nature Conservation

The Washington Convention (Convention on the International Trade in Endangered Species of fauna and flora, or CITES) signed in 1973 classifies the species in three groups or annexes. Annex I contains the species that are threatened by extinction and which are not allowed to be traded; exceptions can be authorized in cases of duly justified research needs. Annex II lists the species that are allowed to be traded under special permits, which have to be issued before their transport. Annex III deals with species that are protected in their countries of origin and their trade is subject to the same conditions as in Annex II.

This text is enforced by the regulation No, 338/97 of the European Commission, which has very similar contents.

In a general way capture, transport, keeping, disposal and the destruction of animals listed in one of the annexes of the Washington Convention are subject to such regulations.

The Keeping of Dangerous Animals

In French texts there is no clear definition of dangerous animals. The decree of 21 November 1997 defines them as animals that "present serious danger or inconvenience". Most often they are compared to non-domestic animals and fall under the corresponding regulations.

It is therefore not permitted to let them run free and contravention can lead to a fine and their being taken away and placed in an appropriate location (pound or in the case of snakes a natural history museum or zoological garden).

There is a duty of protection and of good treatment regarding nutrition, care, accommodation (cage appropriate for the species) and handling, excluding in particular instruments that can cause harm.

Certificate of Competency and Permission to Open an Establishment

The certificate of competency establishes that the person possessing it has the necessary competency to keep non-domestic animals. It is provided by the Directorate of Veterinary Services and can be replaced by an exemption, if the animals are kept privately without presentation, visit or sale to the public.

A permit to open is absolutely necessary for any establishment defined as a farm for non-domestic animals, whatever their number, if the objective is to show, sell or otherwise dispose of individuals from the farm, irrespective of whether they are born there or not. The installations must assure the safety of visitors and of staff as well as the protection of the animals kept in them. The person in charge of the establishment must have a certificate of competency. In addition a register of origin of each of the animals and its movements must be kept up to date and presented when required. The other formalities necessary for any enterprise must also be complied with.

Controls are carried out by sworn agents and commissioners: officials of the judiciary police, customs, the National Forestry Office, the National Hunting Office, or the Fisheries Council. However, visits to private homes can only take place in the case of a judicial enquiry following a complaint or a duly noted contravention.

Bibliography

Angel F., 1950 – *Vie et mœurs des serpents*. Paris, Payot, pp. 319.
Arbonnier M., 2000 – *Arbres, arbustes et lianes des zones sèches d'Afrique de l'Ouest*. Montpellier et Paris. Cirad/MNHN/IUCN, pp. 542.
Audebert F., Sorkine M., Robbe-Vincent A., Bon C., 1994 – Viper bites in France: Clinical and biological evaluation; kinetics of envenomations. *Hum. Exp. Toxicol.* 13: 683-888.
Bailay G. S., 1998 – *Enzymes from snake venom*. Fort Collins, Alaken Inc., pp. 736.
Barbault R., 1971 – Les peuplements d'ophidiens des savanes de Lamto (Côte d'Ivoire). *Ann. Univ. Abidjan,* sér. E, 4: 133-193.
Bauchot R., 1994 – *Les serpents*. Paris, Bordas, pp. 239.
Bellairs A., 1971 – *Les reptiles*. Paris, Bordas, coll. La grande encyclopédie de la nature, vol. X, pp. 767.
Bjarnason J. B., Fox J. W., 1994 – Hemorrhagic metalloproteinases from snake venoms. *Pharmac. Then.,* 62: 325-372.
Bon C., 1992 – Les neurotoxines phospholipases A_2 de venins de serpents. *Ann. IP/actualités,* 3: 45-54.
Bon C., Goyffon M., 1996 – *Envenomings and their treatments*. Lyon, Fond. Marcel Mérieux éd., pp. 343.
Boquet P., 1948 – *Venins de serpents et antivenins*. Paris, Flammarion, coll. de l'Institut Pasteur, pp. 157.
Boujot C., 2001 – *Le venin*. Paris, Stock, coll. Un ordre d'idées, pp. 232.
Boulain J.-C., Menez A., 1982 – Neurotoxin-specific immunoglobulins accelerate dissociation of the neurotoxine-acetylcholine receptor complex. *Science,* 217: 732-733.
Braud S., Wisner A., Bon C., 1999 – Venins de serpent et hémostase. *Ann. IP/actualités,* 10: 195-206.
Bücherl W., Buckley E., Deulofeu V., 1968 – *Venomous animals and their venoms*. New York, Academic Press, vol. I, pp. 707. et vol. II, pp. 687.
Calmette A., 1907 – *Les venins. Les animaux venimeux et la sérothérapie antivenimeuse*. Paris, Masson et Cie, pp. 396.
Chippaux J.-P., 1986 – *Les serpents de la Guyane française*. Paris, Orstom, coll. Faune tropicale XXVII, pp. 165.
Chippaux J.-P., 1998 – Snake-bites: appraisal of the global situation. *Bull. WHO,* 76: 515-524.
Chippaux J.-P., 1999 – L'envenimation ophidienne en Afrique : épidémiologie, clinique et traitement. *Ann. IP/actualités,* 10 : 161-171.

Chippaux J.-P., 2001 – *Les serpents d'Afrique occidentale et centrale.* Paris, IRD, coll. Faune et Flore Tropicale 35, pp. 292.
Chippaux J.-P., Goyffon M., 1989 – Les morsures accidentelles de serpent en France métropolitaine. *Nouvelle Presse médicale,* 18: 794-795.
Chippaux J.-P., Goyffon M., 1998 – Venoms, antivenoms and immunotherapy. Toxicon, 36: 823-846.
Chippaux J.-P., Goyffon M., 2002 – Les envenimations et leur traitement en Afrique. Numéro spécial *Bull. Soc. Path. Exo.,* 95, 129-224.
Chippaux J.-P., Williams V., White J., 1991 – Snake venom variability: methods of study, results and interpretation. *Toxicon,* 29: 1271-1303.
Curran C., Kauffeld C., 1951 – *Les serpents.* Paris, Payot, pp. 275.
Daltry J. C., Wuster W., Thorpe R. S., 1996 – Diet and snake venom evolution. *Nature,* 379: 537-540.
David P., Ineich I., 1999 – Les serpents venimeux du monde : systématique et répartition. *Dumérilia,* 3: 7-499.
Ducancel F., Wery M., Hayashi A. F., Muller B. H., Stöcklin R., Ménez A., 1999 – Les sarafotoxines de venins de serpent. *Ann IP/actualités,* 10: 183-194.
Duellman W. E., 1979 – *The South American herpetofauna: its origin, evolution, and dispersal.* Lawrence, Univ. Kansas print., Monograph Mus. Nat. Hist. n° 9, pp. 485.
Fretey J., 1989 – *Guide des reptiles de France.* Paris, Hatier, pp. 255.
Goin C. J., Goin O. B., Zug G. R., 1978 – *Introduction to herpetology.* San Francisco, W. H. Freeman & Co., pp. 378.
Goyffon M., Chippaux J.-P., 1990 – *Animaux venimeux terrestres.* Encycl. méd.-chir., Paris, Intoxications, 16078 A[10], pp. 14.
Goyffon M., Heurtault J., 1995 – *La fonction venimeuse.* Paris, Masson, pp. 284.
Grassé P.-P., 1970 – *Traité de zoologie.* Paris, Masson, vol. XIV, fasc. 2, pp. 680. et fasc. 3, pp. 1428.
Grigg G., Shine R., Ehmann H., 1985 – *Biology of Australasian frogs and reptiles.* Surrey Beatty & Sons Pty Limited, Chipping Norton, pp. 527.
Grzimek B., 1992 – *Le monde animal. VI. Les Reptiles.* Zurich, Stauffacher S. A. éd., vol. VI, pp. 585.
Harris J. B., 1986 – Natural toxins, Animal, plant, and microbial. Oxford, Clarendon Press, pp. 353.
Harvey A. L., 1991 – *Snake toxins.* New York, Pergamon Press Inc., pp. 460.
Heatwole H. F., Taylor J., 1987 – *Ecology of reptiles.* Chipping Norton, Surrey Beatty & Sons Pty Limited, pp. 325.
Holman J. A., 2000 – *Fossil snakes of North America. Origin, evolution, distribution, paleoecology.* Bloomington, Indiana University Press ed., pp. 357.
Houghton P. J., Osibogun I. M., 1993 – Flowering plants used against snakebite. *J. Ethnopharmacol.,* 39: 1-29.

Bibliography

Hutton R. A., Warrel D. A., 1993 – Action of snake venom components on the haemostatic system. *Blood Rev.,* 7: 176-189.
Ismail M., Aly M. H. M., Abd-Elsalam M. A., Morad A. M., 1996 – A three-compartment open pharmacokinetic model can explain variable toxicities of cobra venoms and their toxins. *Toxicon,* 34: 1011-1026.
Ivanov M., Rage J.-C., Szyndlar Z., Venczel M., 2000 – Histoire et origine des faunes de serpents en Europe. *Bull. Soc. Herp. Fr.,* 96: 15-24.
Kini R. M., Evans H. J., 1990 – Effects of snake venom proteins on blood platelets. *Toxicon,* 28: 1387-1422.
Kochva E., 1987 – The origin of snakes and evolution of the venom apparatus. Toxicon, 25: 65-106.
Lee C. V., 1979 – "Snake venoms". In: *Handbook of experimental pharmacology,* vol. 52, Berlin, Springer-Verlag, pp. 1 130.
Markland F. S. Jr., 1998 – Snake venoms and the hemostatic system. *Toxicon,* 36: 1749-1800.
Martz W., 1992 – Plants with a reputation against snakebite. *Toxicon,* 30: 1131-1142.
Matz G., Weber D., 1998 – *Guide des amphibiens et reptiles d'Europe.* Lausanne, Delachaux & Niestlé, pp. 292.
Meier J., White J., 1995 – *Handbook of clinical toxicology of animal venoms and poisons.* Boca Raton, CRC Press Inc., pp. 752.
Ménez A., 1987 – Les venins de serpents. *La Recherche,* 190: 886-893.
Ménez A., 1993 – Les structures des toxines des animaux venimeux. *Pour la Science,* 190: 34-40.
Ménez A., 2002 – *Perspectives in molecular toxicology.* Chichester, Wiley & Sons, Ltd, pp. 485.
Mion G., Goyffon M., 2000 – *Les envenimations graves.* Paris, Arnette éd., pp. 164.
Morris R., Morris D., 1965 – *Des serpents et des hommes.* Paris, Stock, coll. Les livres de la nature, pp. 224.
Naulleau G., 1997 – *La vipère aspic.* Angoulême, Éveil Nature éd., pp. 72.
Newman E., Hartline P., 1982 – La perception infrarouge des serpents. *Pour la Science,* 55: 92-104.
Ownby C. L., Odell G. V., 1989 – *Natural toxins. Characterization, pharmacology and therapeutics.* Oxford, Pergamon Press, pp. 211.
Parker H. W., Bellairs A., 1971 – *Les amphibiens et les reptiles.* Paris, Bordas, coll. La grande encyclopédie de la nature vol. IX, pp. 383.
Parrish H. M., 1980 – *Poisonous snakebites in the United States.* New York, Vantage Press, pp. 472.
Peters J. A., 1964 – *Dictionary of herpetology.* New York, Hafner Publishing Co., pp. 426.
Phelps T., 1981 – *Poisonous snakes.* Poole, Blandford Press Ltd., pp. 237.
Phisalix M., 1918 – *Animaux venimeux et venins.* Paris, Masson, vol. I, pp. 864, vol. II, pp. 864.

Pirkle F. S., Markland F. S. Jr., 1988 – *Hemostasis and animal venoms*. New York, Marcel Dekker Inc., pp. 628.

Rage J.-C., 1982 – L'histoire des serpents. *Pour la Science*, 54 (avril 1982): 16-27.

Rivière G., Bon C., 1999 – Immunothérapie antivenimeuse des envenimations ophidiennes : vers une approche rationnelle d'un traitement empirique. *Ann. IP/actualités*, 10: 173-182.

Rivière G., Choumet V., Audebert F., Sabouraud A., Debray M., Scherrmann J.-M., Bon C., 1997 – Effect of antivenom on venom pharmacokinetics in experimentally envenomed rabbits: towards an optimization of antivenom therapy. *J. Pharmacol. Exp. Ther.*, 281: 1-8.

Rodda G. H., Sawai Y., Chiszar D., Tanaka H., 1999 – *Problem snake management. The Habu and the Brown Treesnake*. Ithaca, Comstock Publishing Associates, Cornell University Press, pp. 534.

Rosing J., Tans G., 1992 – Structural and functional properties of snake venom prothrombin activators. *Toxicon*, 30: 1515-1527.

Russell F. E., 1980 – *Snake venom poisoning*. Philadelphie, J. B. Lippincott Comp., pp. 562.

Russell F. E., 1988 – Snake venom immunology: historical and practical considerations. *J. Toxicol. Toxin Rev.*, 7: 1-82.

Samama M., 1990 – *Physiologie et exploration de l'hémostase*. Paris, Doin, pp. 233.

Seigel R. A., Collins J. T., Novak S. S., 1987 – *Snakes. Ecology and evolutionary biology*. New York, McGraw-Hill Publishing Company, pp. 529.

Seigel R. A., Hunt L. E., Knight J. L., Malaret L. Zuschlag N. L., 1984 – *Vertebrate ecology and systematics*. Lawrence, Univ of Kansas, pp. 278.

Smith S. C., Brinkous K. M., 1991 – Inventory of exogenous platelet-aggregating agents derived from venoms. *Thromb. Haemost.*, 66: 259-263.

Stocker K., 1994 – Inventory of exogenous hemostatic factors affecting the prothrombin activating pathways. *Thromb. Haemost.*, 71: 257-260.

Stocker K. F., 1990 – *Medical use of snake venoms*. Proteins. Boca Raton, CRC Press, pp. 272.

Sutherland S. K., Coulter A. R., Harris R. D., 1979 – Rationalisation of first-aid measures for elapid snake bite. Lancet, 1: 183-186.

Tipton K. F., Dajas F., 1994 – *Neurotoxins in neurobiology, their action and applications*. New York, Ellis Horwood ed., coll. Neuroscience, pp. 196.

Tu A. T., 1982 – *Rattlesnake venoms, their actions and treatment*. New York, Marcel Dekker, Inc., pp. 393.

Tu A. T., 1991 – *Reptile venoms and toxins*. New York, Marcel Dekker Inc., Handbook of natural toxins, vol. V, pp. 827.

Tu A. T., 1988 – *Venoms: Chemistry and molecular biology*. New York, John Wiley and Sons, pp. 560.

Tun-Pe, Aye-Aye-Myint, Khin-Ei-Han, Thi-Ha, Tin-Nu-Swe, 1995 – Local compression pads as a first-aid measure for victims of bites by Russell's viper (*Daboia russelii siamensis*) in Myanmar. *Trans. R. Soc. Trop. Med. Hyg.*, 89: 293-295.

Bibliography

Visser J., Chapman D. S., 1982 – *Snakes and snakebites. Venomous snakes and management of snakebite in Southern Africa.* Cape Town, Purnell ed., pp. 152.

Weinstein S. A., Kardong K. V., 1994 – Properties of Duvernoy's secretions from opistoglyphous and aglyphous colubrid snakes. *Toxicon,* 32: 1161-1185.

Zingali R. B., Bon C., 1991 – Les protéines de venins de serpents agissant sur les plaquettes sanguines. *Ann. IP/Actualités,* 4: 267-276.

Zug G. R., 1993 – *Herpetology. An introductory biology of amphibians and reptiles.* San Diego, Academic Press, pp. 527.

Wilde H., Thipkong P., Sitprija V., Chaiyabutr N., 1996 – Heterologous antisera and antivenoms are essential biologicals: perspectives on a worldwide crisis. *Ann. Intern. Med.,* 125: 233-236.

Index

Note: Page numbers ending in "f" refer to figures. Page numbers ending in "t" refer to tables.

Acanthophis, 31
acetylcholinesterase, 58, 58f
Acrochordidae, 14
Acrochordoidea, 11, 14
Adenorhinos, 36
adrenergic toxins, 85–88
African pythons, 14
Afronatrix, 181
Agkistrodon, 43, 98, 101–103, 121–123, 174–175
Agkistrodontini, 42–43
agregoserpentins, 69
agricultural activities, 182–184
Aipysurus, 177
Alethinophidians, 4, 10–11, 13, 25t, 26–43
amniotes, 24
amylase, 61
anaconda, 14
analgesic effect, 127–128
anatomy, 18–24
Aniliidae, 10, 13
Anilius scytale, 13
Aniloidea, 11, 13
animal species changes, 160
Anomalepididae, 25
Anomalophiidae, 11
Anomochilus, 13
antagonism, 132
anti-edematous action, 129
anti-inflammatory action, 128–129
anti-necrosis action, 129
antibodies fragmentation, 158–159
antibodies pharmacokinetics, 147–150

antibody quality, 158–160
anticoagulant activity, 102–105
antidote effect, 130–134
antidotes, 125–165
antigen-antibody binding, 146
antigen quality, 157
antiseptic action, 129
antivenin administration, 152–154, 152f, 154f
antivenin availability, 239–242, 241f
antivenin cross-reactivity, 146–147
antivenin indications, 150–152
antivenin production, 138–145
antivenin purification, 139–144, 139t
antivenin quantity, 161
antivenin reactions, 154–156, 155t
antivenin stability, 145
antivenom producers, 257–269
Asian pitviper, 42
aspiration, 134–135
Atheris, 36, 37
Atractaspididae, 16, 30, 53–54
Atractaspis, 30, 54, 68, 90, 198, 224

bamboo pitviper, 42
biological control, 244
birth of snakes, 178–179
Bitis, 16, 36–37, 66, 68, 70, 101, 123, 177, 179, 201–202, 224
blood coagulation, 92, 118–119. *See also* coagulation factors
blood platelets, 105–107. *See also* platelet aggregation
blood pressure, 89–90

Boa, 14
Boidae, 4, 10–14, 51
Boiga, 15, 28–29
Booidae, 10, 11, 13
Bothrops, 41, 68–72, 90, 96–97, 102–103, 105–106, 111, 116, 119, 121–123, 133, 147, 152, 163, 179, 182, 192, 200, 216, 218, 223–224, 243
Boulengerina, 31
Bungarus, 31–32, 64, 66, 133, 160, 164, 206, 207, 235
burrowing adder, 30
bush viper, 37
bushmaster, 42

calciseptin, 67
Calliophis, 17
Calloselasma, 43, 68, 69, 71–72, 98, 102, 122–123, 160, 179, 206–207
capturing snakes, 169–171, 242–245
cardio-tonic action, 130
cardio-toxic effects, 228
cardiovascular complications, 225
cardiovascular system, 89–90
Causinae, 35–43
Causus, 16, 36, 179, 181, 182, 184, 202
cells, 78–80, 79t
Cerastes, 37, 101, 102, 105, 106, 198, 201
chemical control, 244–245
chemical denaturation, 136
chemical inhibition, 236
chemical modification, 134
classifications, 9t, 10–12, 24–43, 25t
Clelia, 29
coagulation factors, 92, 95–102, 96f, 118–119
coagulation inhibition, 103, 105
coagulation test, 220f, 231
coagulation time, 219–220
cobra syndrome, 213–214, 234–235

cobra venom factor (CVF), 70, 124
cobras, x, 16, 33
Coluber, 176
Colubridae, 11–12, 14–15, 26–29, 27f, 51–52
Colubroidea, 11, 12
competition, 130–132
complications, 223–225, 228, 238–239
convulsive syndrome, 213
coral snakes, 17, 33
creatinine-phosphokinase (CPK), 111–112
cross reactions, 146–147
crotalids, 15
Crotalinae, 15, 40–43
Crotalus, 40, 55, 66–69, 71, 98, 106, 111, 121, 146, 159, 175, 200
curare-like syndrome, 213–214
curare-like toxins, 64
Cylindrophis, 13
cytolysis, 217
cytolytic syndrome, 217
cytotoxins, 64–65

Daboia, 36–38, 64, 66, 96, 119, 132, 165, 198, 206–209, 223–224
death adder, 31
Deinagkistrodon, 43, 69, 98, 103, 133, 206
demography, 171–172
denaturation, 136
Dendroaspis, 17, 32, 52, 67, 69–70, 87, 107, 112, 131, 138–139, 162, 179, 182, 202, 211–212
dendropeptin, 70
dendrotoxins, 67
density, 171–174, 173f, 179, 179f
dentition, 5, 19, 19f, 48–51, 49f
deshydrogenases, 61
Dinilysia, 10
disintegrins, 68
Dispholidus, 27, 52, 97

Index

disseminated intravascular coagulopathy (DIC syndrome), 220, 236
diuretic action, 130

ear, 21–22, 21f
Echis, 37–39, 53, 68, 80, 97, 103, 119, 132, 147, 174, 182, 201, 205–206, 209, 219, 221, 224
ecology of snakes, 169–182, 243–244
ectotherm vertebrates, 24
eggs, 24, 178–179
Egyptian cobra, x
Egyptian elapid, x
elapid, x
Elapidae, 12, 16–17, 30–35, 52, 61, 66
ELISA tests, 144, 145f, 151, 161
emergences, 177–179
endothelium, 89
Enterobacter, 110
envenomations
    chronology by elapids, 212f
    chronology by viperids, 215f
    clinic of, 211–227
    complications of, 223–225, 228, 238–239
    epidemiology of, 169–210
    first aid for, 228–230, 229t
    monitoring, 230–232, 231f, 239, 239f
    prevention of, 242–245
    severity of, 216t
    treatment of, 228–246
    see also snakebites
enzymatic activity, 108
enzyme activators, 69–70
enzyme inhibitors, 69–70
enzymes, 57–61
Eristocophis, 36, 68, 80, 123
evolution, 3–12, 5f–8f
eye injuries, 225
eyes, 22

factor V activators, 96–97, 119–120
factor X activators, 119
fangs, xi, 19, 48–51, 49f. *See also* dentition
fasciculins, 67, 88, 213
fer-de-lance, 41
fibrin degradation products (FDP), 94–95, 220
fibrin formation, 93–94, 93f, 94f, 100f
fibrin stabilizing factor (FSF), 94
fibrinogenases, 104t
fibrinolysis, 94–95, 102–103
first aid, 228–230, 229t
folklore, 126
French legislation, 271–273

gangrene, 217, 228
glossary, v–vi
Gloydius, 43, 66, 101, 206

Haasiophis, 10
habu, 42
haemolysis, 223
haemorrhagic syndrome, 163–164
haemostasis
    action on, 129
    mechanisms regulating, 95
    physiology of, 92–95
    problems with, 217–222
    stages of, 91f
    toxic effects on, 91–107
    treatment of, 235–236
    and venoms, 95–105, 96f
handling snakes, 185
hatching of snakes, 178–179
hearing, 21–22
Hemachatus, 32, 225
hemorrhagic syndromes, 217–222, 228
horned viper, 37
human behavior, 182–185
hundred-pace-viper, 43

hunting, 177
hyaluronidase, 59–60
hydrolysis, 89, 105
Hydrophis, 35
Hypnale, 43, 206
hypovalaemic shock, 228

immune stimulation, 133
immunization procedures, 138–139, 157–158
immuno-affinity, 158
immunoglobulins, 140–141, 140f, 146–152, 148f, 149t
immunoglobulins isolation, 158
immunotherapy, 125–165, 232–234, 232t, 234f
immunotherapy future, 157–161
immunotherapy side effects, 154–156, 155t
infection-inducing factors, 110
inflammatory response, 108–110, 109f
injection quantity, 150–151
intramuscular injection, 112–115, 113f

king cobra, 33
krait, 31–32

L-amino-acid-oxidase, 59, 59f
Lachesini, 41
Lachesis, 42, 69, 179
lancehead, 41
Lapparentophis, 10
Iatrogenic factors, 110
LD measurement, 75–78, 76t, 77f, 247–248
legends, ix–xi
Leptotyphlopidae, 25
lytic enzymes, 61

Macrovipera, 36, 39, 123, 198, 201
Madtsoiidae, 10, 11, 13

Malayan pit viper, 43
Malpolon, 29, 52, 174
mambas, 17, 32
mambin, 69
mamushi, 43
Maticora, 17
mating, 178, 178f
maxilla, 48–51, 49f. *See also* dentition
mechanical treatment, 134–136
medicine, traditional, 125–136
metallo-proteases, 60–61
Micruroides, 17
Micrurus, 17, 33, 111
minute snakes, 24–25
moccasin, 43
monoclonal antibodies, 159
Montivipera, 36, 39, 198
movements, 177–179
mulga snake, 34
muscarin-like toxins, 64, 85
muscarine-like syndrome, 212, 234–235
myotoxicity, 110
myotoxins, 67–68, 88

Naja, x, 16, 33, 55, 112, 114, 116, 121, 130–133, 147, 159–160, 179, 182, 198, 201–202, 206–207, 209, 211–212, 217, 225
Natrix, 174
necrosis, 109f, 111–112, 217, 228
necrotizing activity, 107–112, 109f
nerve conductivity, 83–88
nerve growth factor (NGF), 68, 117
nervous signal transmission, 81–83
nervous system, 81–88, 82f, 84f
neuro-muscular synapses, 82–83, 82f
neuro-muscular syndromes, 211–214
neurological problems, 161–162
neurotoxins, 61–70, 62t–63t, 65f, 228

Index

neutralization curves, 143–144, 143f, 147–148, 148f, 154, 154f
nicotine-like toxins, 84–85
night adder, 36
non-conventional transmissible agents (NCTA), 160
Notechis, 17, 33, 66, 97, 98, 121, 209

olfactory senses, 22–24, 23f
Ophidians, 10
Ophiophagus, 33
opisthoglyphous colubrid, 15
opisthoglyphous maxilla, 15
organs, 20–24
origins, 3–17
osteology, 18–19
Osyuranus, 90
Oxybelis, 29
Oxyuranus, 34, 66, 67, 97–98, 119, 164, 209, 225

Pachyophis, 10
Palaeonaja, 16
palaeontology, 3–17
Palaeopheidae, 13
pastoral activities, 182–184
Philodryas, 111
phosphoesterases, 58–59, 59f
phospholipases, 57–58, 57f
phospholipids, 105
physical denaturation, 136
physiognomy, 125
pigmy rattlesnake, 41
plants, 125–134, 128t, 131f, 249–256
platelet activation inhibition, 106–107
platelet aggregation, 92, 105–106
platelets, 105–107
Podophis, 10
poison fangs, xi, 19, 48–51, 49f. *See also* dentition

population composition, 180–182, 181f
population density, 171–174, 173f, 179, 179f
posology, 153–154, 161
postsynaptic neurotoxins, 64, 65f, 228
postsynaptic toxins, 83–86
pre-synaptic facilitating toxins, 87–88
pre-synaptic neurotoxins, 66–67, 228
pre-synaptic paralyzing toxins, 86–87
Proptychophis, 15
proteases, 60–61
protein C activation, 102, 120, 122
prothrombin activators, 97–98
Protobothrops, 42, 68–69, 103, 106, 133, 164, 174, 176, 182, 206–207
Pseudechis, 17, 34, 66
Pseudocerastes, 36, 198
Pseudomonas, 110
Pseudonaja, 17, 34–35, 67, 72, 98, 209
Python, 14, 179

radio-immuno assay (RIA), 151
rattlesnake, 40
repellent effect, 126
repellents, 244–245
reproduction, 20–21
Rhabdophis, 28, 97, 174, 176
ringhals, 32
Russelophiidae, 11
Russel's viper, 37–38

sarafotoxins, 68
saw-scaled viper, 38–39
Scolecophidians, 4, 10, 13, 24–25, 25t, 51
sea snakes, 35
Securidaca, 131–132, 131f, 245
senses, 21–24
serin-proteases, 60
serum therapy, xii
shock, 228

Simoliopheidae, 10
Simoliophis, 10
Sistrurus, 15, 41, 137
snakebites
  causes of, vii
  characteristics of, 189–191
  circumstances of, 187–189
  deaths from, 187, 190–209
  diagnosis of, 230–232, 231f
  epidemiology of, 185–210
  first aid for, 228–230, 229t
  incidence of, 185–186, 190–209, 190f, 191f, 193t–195t, 196f, 197f
  lethality of, 187, 189–209
  medical response to, 209–210, 210f
  monitoring, 230–232, 231f, 239, 239f
  morbidity of, 186–187, 189–209, 196f, 199f
  mortality of, 187, 190–209
  number of, vii
  protection against, 245
  reasons for, 187–189
  serum for, xii
  severity of, 191–193, 193t–195t
  studying, 185–187
  treatment of, 160–164, 209–210, 210f, 228–246
  see also envenomations
snakes
  birth of, 178–179
  capture of, 169–171, 242–245
  characteristics of, ix
  classifications of, 9t, 10–12, 24–43, 25t
  control of, 242–245
  demography of, 171–172
  ecology of, 169–182, 243–244
  emergences of, 177–179
  evolution of, 3–12, 5f–8f
  handling, 185
  legends about, ix–xi
  movements of, 177–179
  organs of, 20–24
  origins of, 3–17
  population composition of, 180–182, 181f
  population density of, 171–174, 173f, 179, 179f
  species of, vii
  symbolism of, ix–xi
  territory of, 174–177, 176f, 243
  zoology of, 1–43
  see also snakebites
solenoglyphous maxilla, 16
species numbers, vii
spitting cobra, 32
spitting snakes, 32, 225
Squamata, 24
Strophantus, 132
substitution therapy, 235–236
symptomatic effect, 126–130, 127t, 128t
systematics, 24–43
systemic antidotes, 130–133

Tachymenis, 28
taipan, 34
teeth, 5, 19, 19f, 48–51, 49f
temperature, 24, 177
territory, 174–177, 176f, 243
Thamnophis, 173, 175
Thelotornis, 28, 52, 97, 98
thermoregulation, 177
thermosensitive receptors, 23–24, 24f
thrombin formation, 92, 93f
thrombin-like enzymes, 98–102, 99t–100t
thrombo-elastogramme, 221, 222f
thrombocytes, 69
thrombolectins, 69
tiger snake, 33
tourniquet, 134
toxicity

Index

on cardiovascular system, 89–90
on cells, 78–80, 79t
factors influencing, 75–78, 76t
on haemostasis, 91–107, 91f
mean toxicity, 76–78, 77f
measuring, 75–78, 76t, 77f, 247–248
on nervous system, 81–88, 82f, 84f
on platelets, 105–107
Toxicodryas, 29
toxicokinetics, 112–115
toxicology, 75–124
toxins, 61–70, 62t–63t, 65f
traditional medicine, 125–136
transaminases, 61
transfusions, 235–236
treatment mechanisms, 134–136
treatment protocols, 160–161
treatment strategies, 161–164, 209–210, 210f, 228–246
Trimeresurus, 42, 68–69, 97, 102–103, 122, 133, 179, 207
Tropidolaemus, 69, 106
Tropidopheidae, 13
Tropidophiidae, 13
tumor necrosis factor (TNF), 110
Typhlopidae, 4, 25
Typhlops, 13

vaccinations, 164–165
venom
  antidotes for, 125–165
  chemical inhibition of, 236
  composition of, 54–73
  compounds in, 68–70
  in diagnostics, 118–120
  diffusion of, 113–115, 114f, 229–230, 230f
  discovery of, xi
  enzymes of, 57–61
  identification of, 225–227, 226t
  injected venom, 150–151
  in research, 115–124
  spraying of, 225
  synthesis of, 54–55
  therapeutic uses of, 121–124
  titration of, 225–227, 226t
  toxicity of, 56t, 75–124
  toxicology of, 75–124
  toxins in, 61–70, 62t–63t, 65f
  variability of, 70–73
venom apparatus, 47–73
venom gland, 48–51
venomous animal ownership, 271–273
venomous species, vii
viper syndrome, 214–217, 237–239
Vipera, 15, 39–40, 66, 96, 160, 173, 175, 177–178, 182, 196, 206
Viperidae, 12, 15–16, 35–43, 53, 57, 60, 66
Viperinae, 15, 36–40, 36f
viperine syndrome, 108–111
vipers, 12, 15–16
visceral complications, 223–225, 228
vision, 22
vitamin K deficiencies, 119–120

Walterinnesia, x, 198, 201, 211
water cobra, 31

Xenopeltidae, 13